KOREA
THE TIGER ECONOMY

By Trenholme J Griffin

Published by
Euromoney Publications
in association with:

Banque Indosuez/WI Carr

Daewoo Securities Company Limited

Dongsuh Securities Company Limited

Korea Foreign Trade Association

The Lucky Securities Company Limited

Merrill Lynch Capital Markets

Published by
Euromoney Publications PLC
Nestor House
Playhouse Yard
London EC4V 5EX

Copyright ©
Euromoney Publications PLC
1988

ISBN 1 870031 37 7

All rights reserved.
No part of this book
may be reproduced
in any form or by
any means without
permission from the
publisher

*Typeset and artwork by
Multiplex techniques ltd*

*Printed in Korea by
Pyung Hwa Dang/Hyondae Trading*

Contents

Introduction 1

Chapter I: History and culture of Korea 5

    The Hermit Kingdom 5
        Pre-history 5
        The kingdoms 5
        Japan's colonial attempts 5
        The division of Korea 6
        The Korean War 6
        Re-building Korea 7
    Korea's culture 9
    Religion 9

Chapter II: Geography and people 11

    Climate 11
    Seoul 11
    Provinces, special cities and industrial estates 12
    Regional rivalries 13
    North Korea 14
    People 15

Chapter III: The structure and role of government 19

    The role of government 19
    Structure of government 19
    The principal ministries and agencies 20
    Dealing with government 21
    Legal system 23
    Economic policy formulation 23

Chapter IV: Foreign relations 25

    Relations with the United States 25
    Relations with Japan 26
    Relations with the Union of Soviet Socialist Republics 26
    Relations with the People's Republic of China 26

Chapter V: Domestic politics 29

    Roh Tae Woo and the Democratic Justice Party 29
    The two Kims 30
    Ex-president Chun Doo Hwan 32
    Other political power groups 32

Chapter VI: The economy 35

    Exports and the three blessings 35
    Korea's industrial sectors 35
        Manufacturing 35
        Other industries and resources 42
    Plans for economic diversification 50
    Economic forecasts 50
        Near-term economic prospects 50

Chapter VII: Economic projections 57

    Medium- and long-term prospects 58
    Recent economic developments 59
    The first five-year economic development plan 60
    The second five-year economic development plan 61
    The third five-year economic development plan 61
    The fourth five-year economic development plan 61
    The fifth five-year economic development plan 62
    The sixth five-year economic development plan 63

Chapter VIII: Financial institutions 63

    Commercial banks 63
        City banks 63
        Local banks 64
        Foreign bank branches 65
    The central and special banks 66
    Merchant banks 67
    Other financial institutions 69
        Leasing firms 69
        Insurance firms 69

Chapter IX: The money markets 71

    The formal money markets 71
    The informal money markets 71
    The stock markets 72
        The Korea Stock Exchange 72
        Securities trading 73
        Internationalization 74
        Listing 77
        The over-the-counter-market 77
        Brokerage firms 77
        Securities firms 78
        International issues 78
    The bond market 79

Foreign direct investment 81
Monetary policy 82
Offshore debt financing 83
Public finance 86
Taxation 87
Foreign exchange controls 87

Chapter X: Ground rules for doing business 93

Choosing your business partner 93
Business cards 93
Social relationships 94
The chaebol 96
Non-chaebol partners 98
The decision-making structure 99
Basic standards of decorum and business etiquette 100
The primacy of cash 102
Korean auditing and accounting practices 104
Korean negotiating style 104
The fluidity of contractual agreements 105
Resolving disputes 107
Communications 108
Investigate production capability and monitor quality standards 108
Employee selection 109
    Foreign managers and technicians 109
    Korean employees 110
Technology transfer 112
Major policies and objectives 113
Korea's unique qualities 114
    Major differences between Korean and Japan 115
Deceptive outward appearances 116
Immigration 117

Customs and Excise 117
Currency 120
Living conditions for foreigners 120
Medical facilities and emergencies 120

Chapter XI: Business structures for foreign companies 123

Importing products 123
Localization and standards 124
Technology licensing 125
Informal joint venture 126
Joint venture or subsidiary 126
Choice of business entity 128
    Liaison office 128
    Branch office 128
    Corporation 129
Business case studies involving foreign companies 130
    Two food products companies 130
    Two athletic shoe companies 130
    Two computer companies 130
    Dow chemical company 131
    A Japanese automobile manufacturer 131
    Two US trading companies 131
    Two food manufacturing companies 132
Conclusions 132

Appendices 133

Information for visitors 133
Business glossary and acronyms 137
Korean government offices 139
Economic and trade associations 140
KOTRA overseas offices 143
Foreign embassies in Korea 150

A typical Korean Buddhist temple is made up of many shrines which are often housed in several special halls

Introduction

The Republic of Korea* has one of the fastest growing and most dynamic economies in the world. The phenomenal economic performance of Korea has led commentators to classify it as one of the four *new tigers* of Asia together with Singapore, Taiwan and Hong Kong. Each of these new tigers plays an increasingly important role in the world economy. Although the Korean economic tiger shares many characteristics with its Asian cousins, it is unique in many important respects.

This book is intended to familiarize foreign businessmen and bankers with the Korean business climate as well as to help make their time in Korea more productive and pleasant. The information is intentionally selective rather than exhaustive in order to serve as a useful and manageable tool for foreign businessmen and bankers. The ground rules section of this book in Chapter X focuses on the dangers and pitfalls which may arise when doing business in Korea, emphasizing practical advice and identifying the aspects relevant to doing business in Korea.

In order to be useful, the discussion which follows must by its nature contain broad generalizations. Such generalizations are only useful in describing the characteristics of cultures or other groups rather than individuals. Exceptions to the general rules will inevitably be encountered while doing business in Korea. Furthermore, the attitudes of each generation of Koreans will vary.

It should be emphasized that the material in this book is not intended to be a criticism of either Korea or its citizens, indeed, the intention is to identify differences in an effort to reduce conflicts and misunderstandings and to make business transactions profitable for all parties concerned. Although there is some discussion of the problems associated with doing business in Korea, this is only meant to provide useful information for the foreign businessman and to help him avoid common mistakes.

Not all of the discussion in this book will be applicable to each foreign businessmen or banker. The problems encountered by exporters differ from those of importers; a merchant banker and an insurance executive have different business concerns; and companies licensing technology to Korea will not be subject to the same considerations as joint venture participants. However, certain ground rules and aspects of the business environment are applicable to nearly all foreign businessmen and bankers and these are emphasized. In addition, a banker or insurance executive can learn from the problems encountered in a joint venture and vice-versa.

The transcription of one language to another poses problems which the author has sought to avoid by following the McCune-Reischauer system. However, if a person or organization typically uses another spelling, then that version is substituted. The only real solution to the romanization problem is to learn the Korean alphabet and Korean writing system. This is not particularly difficult and well worth the effort.

Doing business in Korea is a difficult, time-consuming and complex undertaking. Despite the problems and pitfalls which may arise, numerous lucrative business opportunities exist. East Asia is the world's greatest growth area and a significant portion of that growth will occur in Korea. The current trade surplus and high domestic savings rate will make Korea a major factor in international capital and money markets in a few short years. The opportunities to export products from Korea are well proven. In addition, over 42 million increasingly affluent Korean consumers constitute an attractive market for both importers and investors in domestic enterprises. Due to the excellent working habits, relatively low wages, high education level and high productivity of the Korean work force, any foreign company or financial institution whose competitor is doing business in Korea will be at a serious disadvantage. Numerous companies and financial institutions from the US, Japan and Europe are making the commitment necessary to do business in Korea and as a result will be extremely competitive in world markets. Companies which decide not to become involved in business in Korea or similar countries because they believe it is too difficult will face near-certain extinction. In other words, even if you decide not to make the effort to do business in Korea, your competitors are likely to be doing so and you may be out of business or at least substantially less profitable as a result.

Both novice and experienced foreign businessmen and bankers will certainly make mistakes in doing business in Korea. However, with diligence, persistence and hard work, success can be achieved.

---

* The Republic of Korea will be referred to throughout this book as *Korea. For a discussion of the division of Korea see Chapter II.*

The Chamsil Olympic stadium has a total floor space of 132,000 square metres and can accommodate 100,000 spectators

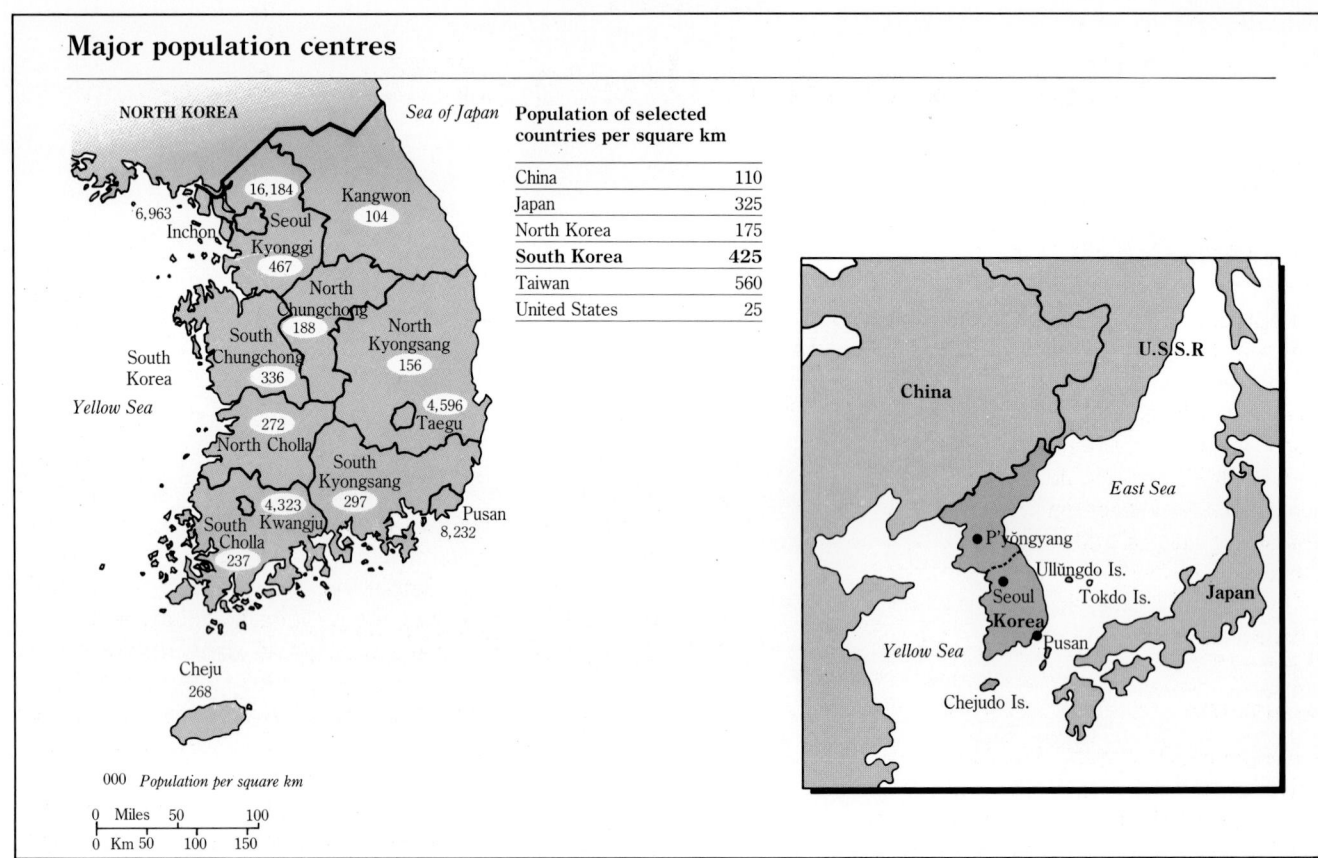

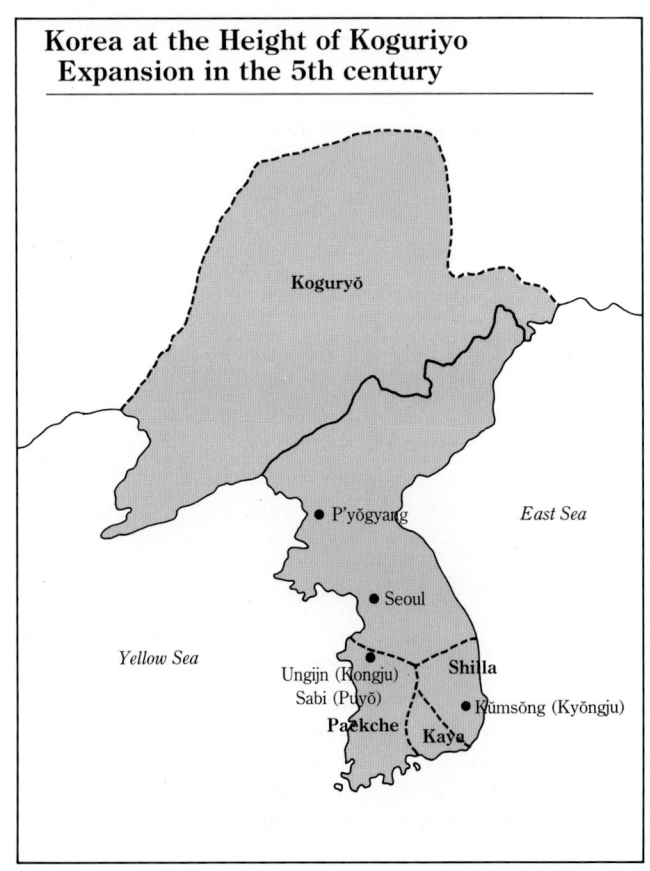

Chapter I
History and culture of Korea

The Hermit Kingdom

Korea was known as the Hermit Kingdom for much of its history because of its high degree of physical and cultural isolation from foreign countries. Korea is a peninsula attached to northwest Asia and runs in a southerly direction for about 1,000 kilometres, approximately the size of Portugal. Only at its extreme north does the Korean peninsula come into contact with foreign countries. The Korean people have a strong cultural and national identity stretching over 5,000 years. For much of its history, Korea has been dominated by China and Japan since the Korean peninsula serves as a natural link between these two great powers.

Pre-history

The Korean people are the descendents of nomadic tribes which originated in Mongolia and migrated to the Korean peninsula around the third millennium BC. Their language is Ural-Altaic and can be traced to an origin in Central Asia. The language is most closely related to Turkish, Hungarian and Finnish. Koreans are a homogenous race, distinct from the Chinese and Japanese. Korean folklore traces the race to a union between the son of the Divine Creator, who came to the earth to live as a human, and a woman from a bear totem tribe. The son of the union was known as Tangun, Korea's first king. While proving the authenticity of this myth is not possible, it is certain that a group known as the *Tung-i* or *eastern bowmen* were the first Koreans to reside on the peninsula. Archaeological evidence suggests an early tribal culture influenced greatly by the Chinese. The Chinese Han Dynasty (202 BC – 220 AD) was successful in colonizing the Korean peninsula during a major portion of this period. The principal city of Korea at that time was located at Lolang, near present day Pyongyang, the capital of north Korea.

The kingdoms

The first of Korea's kingdoms emerged around 100 BC and was known as the Koguryo Kingdom. This kingdom ruled part of present day Manchuria and much of the northern half of the Korean peninsula. The Paekche Kingdom was formed in approximately 250 AD in southwest Korea. About 100 years later the Shilla Kingdom emerged in southeast Korea. The three kingdoms fought almost continually for over 700 years for control of Korea. At first Koguryo was dominant, but later fell into decline, largely as a result of the tremendous burden of fighting the neighbouring Chinese. The Korean peninsula was first unified in 668 AD by the Kim Dynasty of the Shilla Kingdom when it managed to defeat both Koguryo and Paekche. Buddhist culture flourished under this reign. The capital of Shilla was Kyungju, a city in which rich historical relics have been preserved to this day.

Eventually the excesses of the rulers of Shilla resulted in a revolt lead by Wang Kon, a general in a rebel army. In 936, Shilla fell and the Koryo Kingdom was born. The Koryo dynasty lasted for 475 years, during which Korean art and culture flourished and Buddhist monks attained great power in the Koryo court. Despite the advances made during the reign of the Koryo kings, the Kingdom again fell into decline due to infighting and pressures from Mongolian invaders. The Yi Dynasty was established in 1392 by General Yi Sung Kye after he deposed the Shilla king and moved the capital to Seoul. The Yi Dynasty continued for over 500 years and was primarily responsible for the rise of Confucianism in Korean culture. A strong civil service system organized according to Confucian tenets was established by the kings who ruled during this period. Part of the reason for the rise of Confucianism was the desire of Yi Dynasty kings and their ministers to lessen the power of the Buddhist monks. Under the rule of King Sejong, Korean culture made particular progress; for example, a royally appointed group of scholars devised the Korean phonetic writing system. Japanese and Manchu invaders repeatedly ravaged Korea during the late 15th, 16th and 17th centuries. One particularly savage invasion occurred when Japanese forces led by Hideyoshi invaded Korea in 1592.

Western contact with Korea was initiated by traders and missionaries around 1600 AD. However, the conflict between western culture and traditional Korean teachings and values created significant impediments to meaningful interchange with often tragic results. Large numbers of Christian converts and some foreign missionaries were executed during various campaigns against *barbarian* foreign influences.

Japan's colonial attempts

Unlike Japan and China, Korea's commercial trade with the outside world occurred only relatively recently. After forcibly obtaining a trade treaty, Japan began a calculated campaign to colonize Korea. These efforts were furthered by Japan's defeat of China in the Sino-Japanese War in 1894-95, and Japan's victory in the Russo-Japanese War of 1904-05. Japan's treachery included engineering the assassination of

Queen Min of Korea, and forcing the abdication of King Kojong in favour of his feeble son (who was in turn forced to marry a Japanese woman and given a Japanese title). Japan increased its grip on Korea until outright annexation occurred in 1910. During their occupation of Korea, the Japanese pursued a ruthless policy of cultural suppresion in which the Japanese culture was forced upon the Korean people. Japanese was the only language taught in schools and Korean history was not a part of the curriculum. Many Koreans were coerced into adopting Japanese names and the Shinto religion. Korean men were also conscripted into the Japanese armed forces.

During this period of colonization, Japan made no attempt to industrialize or train the Korean people. While infrastructure in Korea was marginally improved during the Japanese occupation, the economy remained largely agrarian and subservient to the needs of Japan. Korea's resources were exploited by the Japanese without concern for their proper management over the long term. To make matters worse, the high taxes imposed by the Japanese government left little capital for investment in the economy. Japanese citizens would eventually expropriate ownership of nearly 80% of Korea's rice fields.

Attempts to create an independent Korea continued during the Japanese occupation. A major impetus to the Korean independence movement was US President Wilson's call for the right of "self determination" for weaker nations at the Paris Peace Conference. A provisional Korean government was organized in Shanghai to fight for independence. The efforts to obtain Korea's independence reached their peak in connection with the March First Independence movement of 1919. The movement began when a group of Korean patriots drafted and proclaimed Korea's independence at a place in Seoul today called *Pagoda Park* by foreigners. Peaceful demonstrations and civil disobedience in support of independence were brutally dealt with by the Japanese. Japan's imperialism continued in Korea until its defeat by the Allied forces in the Second World War.

The division of Korea

After the USSR entered the war against Japan, the Allied forces agreed that Soviet troops would occupy Korea north of the 38th parallel and American troops would occupy the southern portion of Korea. In theory, the occupation was intended to be a temporary arrangement lasting only until elections supervised by the UN could be held. The USSR quickly set about organizing a communist government in the north. Kim Il Sung quickly consolidated power with their help and began his dictatorship which still continues. The ruthless activities associated with establishing the communist regime caused a flood of refugees to flow south to US-occupied Korea.

Following a period of political strife, elections supervised by the UN were finally held in the south in 1948 in which a National Assembly and President Syngman Rhee were elected. The Republic of Korea (south Korea) was formed on August 15 1948, the third anniversary of Korea's liberation. The People's Republic of Korea (north Korea) headed by General Secretary Kim Il Sung was established in September of that same year. The communists, with the support of their Soviet allies, stepped up a guerrilla and propaganda campaign to destabilize the south after the formation of the Republic of Korea. Because it was literally starting with nothing, the government of south Korea faced a formidable task in rebuilding the country. Most of the few existing industrial facilities had been built by the Japanese in the north. As a result the agrarian south had no significant industrial base upon which to build an economy. In addition, the huge number of refugees from the north were a considerable burden upon the financially constrained government of the Republic of Korea.

The Korean War

After the respective governments in the north and south were formed, both the US and USSR reduced their physical presence in their occupied sectors. However, the USSR devoted considerable resources to building the north's military capacity. By early 1950, north Korea held a very large military advantage. The combination of popular dissatisfaction with the Syngman Rhee government and the unfortunate statement by US Secretary of State Dean Acheson that Korea was outside America's defence perimeter in Asia, caused the leaders of north Korea to believe a war to unite the country could be won. The communists decided to use their military advantage to stage a surprise attack on south Korea. On the morning of June 25 1950, without a declaration of war, north Korea attacked. The assault was lead by Russian-built tanks and was only halted at the Naktong River near Taegu. Eventually, only a small perimeter around the southern port city of Pusan was held by south Korea and a few US troops.

President Truman immediately ordered the US military to assist south Korea and requested the assistance of the UN. Fortunately, the USSR's boycott of the UN Security Council allowed the Security Council to pass a resolution demanding that the communists withdraw to the original border. When it was clear that the communists did not intend to comply, troops from 16 other countries from the UN joined the Korean and US armed forces in the fight against north Korea. US Army General Douglas MacArthur was placed in command of the UN Combined Forces. His surprise amphibious landing at Inchon near Seoul struck the north Korean Army at its flank and severed supply lines. The UN forces quickly pressed north along the peninsula almost as far as the Yalu river which marks Korea's border with the

Under Confucian philosophy education is the only real way for an individual to increase his/her status in Korean society

Under Confucian teachings, an inferior is expected to be obedient to his superior and the superior to be benevolent to his inferior

People's Republic of China. To prevent the defeat of north Korea, Chinese Communist volunteers intervened in huge numbers and pushed the UN forces back across the 38th parallel. Inconclusive but fierce fighting continued for two years while truce negotiations were being held. On July 27 1953, an Armistice Agreement was signed which created a demilitarized zone four kilometres wide from one coast of Korea to the other along the 38th parallel. A military Armistice Commission and a Neutral Nations Supervisory Commission continue to supervise the truce.

Re-building Korea

The Korean War left the territory controlled by south Korea in ruins. The city of Seoul was particularly devastated since it had changed hands four times during the course of the war. Little in the city was left standing. Over 60% of south Korea's industrial facilities were destroyed. The damage from the war was estimated to be US$3 billion, equal to the nation's combined GNP for 1952 and 1953. The war also caused enormous social dislocation, as 225,784 Korean soldiers died in combat and approximately 380,000 Korean civilians were killed in war-related actions or murdered by north Korean troops and secret police. Families were separated and scattered throughout Korea. Countless orphans, widows and severely injured veterans were left to pick up the pieces of the shattered society.

In response to unethical and illegal political manoeuvering by President Rhee and his Liberal Party to maintain their grip on power, students and other dissatisfied individuals organized anti-government demonstrations. Particularly severe and widespread demonstrations took place on April 19 1960. The demonstrations continued until April 25 when the military refused to intervene to stop a major demonstration. President Rhee was force to resign on April 27. Shortly afterward, President Rhee moved to Hawaii where he died in exile.

The Second Korean Republic was formed by an interim government after President Rhee's resignation. A European-style cabinet form of government was adopted with Chang Myon as the prime minister. Yun Po Son was selected to serve as the figurehead president of Korea. The Second Republic lasted only nine months. The government of prime minister Chang Myon was ineffective largely because of the political bickering which followed President Rhee's resignation. In response to the ineffectiveness of the Chang Myon government, and because of deteriorating economic conditions, a coup d'etat was staged on May 16 1961 by a group of military officers headed by General Park Chung Hee. Chang Myon was deposed by Yun Po Son but was allowed to retain his position for ten months. The military junta continued in power until 1963, when General Park retired from the military and was elected to head a civilian government. The focus of the Park government was to eliminate corruption and evil practices in government. President Park was later narrowly re-elected in 1967 and 1971. In 1972, the Park government revised the Constitution in a manner which curtailed the powers of the National Assembly and the judiciary and dramatically increased the power of the president.

Despite the criticism levelled against the Park government on political grounds, its economic policies were a major success. The Korean economy under the leadership of the Park government continued to perform well until 1979. During that year, increasing unemployment and inflation caused by both internal and external factors resulted in a wave of anti-government demonstrations. President Park responded by imposing martial law and other measures intended to suppress dissent. In October 1979, President Park was assassinated by Kim Jae Gyu, the head of the Korean Central Intelligence Agency. A short-lived interim government headed by Choi Kyo Hak followed Park's death. Further economic and political turmoil occurred during this period. On December 12 1980, a group of junior generals, principally from the 11th graduating class of the Korean Military Academy headed by Major General Chun Doo Hwan, seized control of the Korean military. Approximately 40 senior generals were forced to retire. The young generals consolidated their power for several months and on May 17 1980 took control of the government. The military government immediately launched an anti-corruption drive which was very popular due to excesses which occurred during the last years of the Park government. However, increasing domestic unrest resulted in the imposition of martial law in 1980.

General Chun resigned from the army and was elected President of the Fifth Korean Republic in 1981 under a new Constitution. Chun Doo Hwan was inaugurated as the nation's twelfth president on March 3 1981, and served a seven-year term. The economic policies of the Chun government were very successful, most notably the increase in the country's per capita income to over US$2850. Opposition forces claim this economic progress was achieved through political repression but most neutral observers view Chun's role as important in moving Korea toward its goal of joining the ranks of democratic advanced countries.

President Roh Tae Woo was elected to succeed President Chun on December 17 1987 and took office on February 25 1988. This was probably the first peaceful transition of power in Korea's 5,000-year history. A new constitution took effect on the same day as Roh's inauguration creating the Sixth Korean Republic. While this new constitution gives the National Assembly a more significant role, President Roh will continue to have the lion's share of power in Korea. Opposition forces, led by Kim Young Sam, Kim Dae Jung and others, will continue to challenge the government of President Roh. A full discussion of the current political situation in Korea will be found in Chapter IV.

Korea's culture

Korea is a country built on relationships. All social and business interaction in Korea is structured on the basis of the relationships of the parties involved. It is important to realize that nearly every relationship in Korea is based on one person being considered superior to another. The list of factors to be taken into consideration in determining which person has superior status is so extensive that there is nearly always something which distinguishes any two given individuals. Even in the case of twins, the first born will have superior status. While the determination of relevant status is complex, any Korean will immediately sense his or her place in the social hierarchy.

The emphasis on relationships in Korea arises from a set of moral teachings known as Confucianism, which have been the primary influence on the behaviour and customs of the Korean people. While not technically a religion, Confucianism has provided the Korean people with rules which define moral and appropriate behaviour and the proper relationship between the government and its citizens. Confucian society is defined in terms of the five relationships: between father and son there should be affection; between ruler and minister, righteousness; between husband and wife, attention to separate functions; between old and young, proper order and respect; and between friends, faithfulness.

The Confucian philosophy teaches that each person has his place in a rational hierarchical social order and that the preservation of harmony within this social order is of paramount importance. Under Confucian teachings, an inferior is expected to be obedient to his superior and the superior to be benevolent to his inferior. In practice, the obedience component is emphasized over the benevolence component in order to maintain the status quo. Furthermore, the emphasis on preserving harmony can result in a lack of mobility between levels in the hierarchy. Education is the only real way for an individual to increase his status in Korean society. It is therefore not surprising that such a premium is placed on education in Korea as well as the other countries in Confucian East Asia (Japan, China, Taiwan and Hong Kong).

Religion

The tribal communities which developed into the kingdoms ruling Korea by the first century BC incorporated many animist rituals which served to bind the communities together. A new chieftain or king would pray at his forebear's altar or shrine which would become more important as that leader's fortunes grew. The shrines became the focus of each kingdom and supported officials and priests who performed the ceremonies. Although more sophisticated religions, such as *Taoism, Buddhism* and *Confucianism,* entered Korea from China during the *Three Kingdoms* era, the older beliefs did not immediately die out. Nothing in the new religions opposed the rituals and rites of animism (or each other), and the people were able to evolve a personal, tolerant and harmonious spirit, the *Hwarang.*

In the late nineteenth and early twentieth centuries, many of the ancient rituals were revived in order to express the growing sense of identity and nationalism. Some of the modern sects and religions, such as the *Ch'ongdo* and *Taejonggyo* claim the mantle of that ancient spirit of Hwarang.

Buddhism

In its earliest form, Buddhism was a set of premises that offered an escape from the endless Hindu cycle of rebirth and reincarnation. However, one of its strengths was the ability to absorb and incorporate local superstitions as it spread from place to place, reaching Korea in the fourth century AD. So rapidly and deeply did it root in Korea that by the sixth century, priests, scriptures and artisans were being sent to Japan to form the basis of their early Buddhist culture. Koreans became renowned for the purity and dedication of their worship.

In the seventh century, when the Shilla kingdom unified the country, Buddhism became the state religion and its influence permeated every level of decision-making. Kyongju, the capital of the Shilla kingdom is virtually an open-air museum of Buddhist relics and art objects. However, the power that the priests were able to wield (and the inevitable corruption of some of them) made them the target of blame for the thirteenth century Mongol invasions. This was the end of Buddhism as a state religion, although it continued to flourish as a personal one. It has the largest following of any of Korea's religions with 7.5 million believers (16% of the population).

Confucianism

Confucianism in Korea is a system of education, ceremony and civil administration. It is not technically a religion, but rather an ethical and moral system explicitly regulating the proper conduct of social relationships. The *five relationships* teach that: between father and son there should be affection; between ruler and minister, righteousness; between husband and wife, attention to separate functions; between old and young, proper order and respect; and between friends, faithfulness. Arriving in Korea at about the same time as Buddhism, the two religions complemented each other by attending to the social and the personal relationships.

After the Mongol invasions, Confucianism became the guiding precept for the state and presided over considerable progress in social reform, modernization and the development of judicial systems during the fifteenth and sixteenth centuries. Establishing nationwide examinations for the bureaucracy helped to ensure that the government was meritocratic and that political leaders were also the intellectual leaders.

The legacy of Confucianism is still an important feature of Korean life, expressed by reverence for learning and culture, social stability and respect for the past. Ancestral rites and memorial ceremonies are still held regularly and there are over 200 *hyanggyo*, or Confucian academies, attempting to make Confucian values more relevant to a modern, industrial society.

Christianity

Catholicism entered Korea through China as early as the seventeenth century through copies of missionaries' translations, but the first priests only crossed the border in 1785. The early church was a secretive underground movement subject to bouts of persecution and martyrdom. In 1876 Korea's cultural and religious isolation came to an end and the powers forced Korea to sign treaties guaranteeing the safety of missionaries.

Protestant missionaries arriving in Korea after the treaties fared much better than the Catholics, bringing education and advanced technology with them to serve the people. The missionaries arranged for able converts to continue their education abroad and helped to support the independence movement in the early twentieth century. Membership in the Protestant churches is very rapidly growing in Korea.

Expectation of life at birth			
	Year or Period	Male (years)	Female (years)
Japan	**1986**	**75.23**	**80.93**
Iceland	193–1984	73.96	80.20
Sweden	1984	73.84	79.89
Norway	1982–1983	72.69	79.54
Australia	1984	72.59	79.09
U.S.A.[a]	1983	71.60	78.80
France	1982	70.73	78.85
U.K.[b]	1982–1984	71.60	77.60
Germany, F.R.	1982–1984	70.84	77.47
U.S.S.R.	1975–1980	65.00	74.30
Korea, Rep. of	1980–1985	62.70	66.60

a) White only. b) England and Wales only.
Source: Ministry of Health and Welfare, Japan; U.N., *Demographic Yearbook*, 1985

Chapter II
Geography and people

Korea is a mountainous peninsula jutting from northeast Asia toward the south for approximately 1,000 kilometres. The nation is geographically separated from Manchuria by the Amnok (Yalu) and Tuman rivers and Mt Paektu. The former river discharges in the Yellow Sea and the latter river into the Eastern Sea (the final 16 kilometres of this river mark Korea's boundary with the USSR). At its closest point, the Korean peninsula is 216 kilometres from Japan. The area occupied by the Republic of Korea is approximately 38,000 square miles (97,000 square kilometres).

Korea has many mountains and hills which take up nearly 70% of its territory. Most of the arable land is located in basins between mountains or in river valleys. Slopes on the east coast are generally steeper than in the west. The southern and western coastlines are highly indented and have an abundance of islands. The east coast has few islands and only a few good ports. Ports on the west coast are more numerous, but are handicapped by a very large tidal range. Most of Korea's major port cities are located on the south coast which has good port locations and a normal tidal range. These southern ports include Ulsan, Pusan, Chinhae, Masan, Yosu and Mokpo. Most rivers run south or west and are generally short and shallow.

Climate

Korea has an east-asian monsoonal climate which results in hot, humid summers and relatively dry, but bitterly cold winters. Winter weather is dominated by a high pressure zone produced by intense cold in Siberia. Snow and cold winds can be expected from late December to March. Winds during late spring and early summer can be gusty and sometimes yellow with dust from the Mongolian Desert. The summer is very hot and humid. Heavy rains will occur sporadically during late June, July and August. The worst of the summer heat and humidity usually occurs during the first three weeks of August. Many foreigners leave Korea during this period. Koreans typically take a short vacation (three or four days) in late August. Fall is the best time to be in Korea since the temperature is pleasant and rain sparse. Mean temperatures are 10.8 C (51.5 F) in spring; 23.6 C (74.5 F) in summer; 13.6 C (56.5 F) in autumn and -1.7 C (28.94 F) in winter.

Seoul

The concentration of decision-making power in Korea has produced an incredible geographical phenomenon. Seoul has become one of the largest cities in the world and the home to nearly one-quarter of the country's population, drawn from the provinces as if by a powerful magnet. The leading universities and cultural institutions are located here along with the head offices of nearly all Korean and foreign companies. For military security reasons, production facilities are largely based in the southern part of the peninsula, but even so, major business decisions are made in Seoul.

Seoul is the heart, soul and brain of Korea. The capital city and its immediate environs are the home of over 11 million of Korea's population of 42 million citizens. This makes Seoul one of the ten largest cities in the world. Seoul's population is continuing to grow at an incredible pace. It has been reported that 20,000 Korean citizens moved to the Seoul metropolitan area from other parts of the country during each month in 1987. Much of the population growth is caused by the fact that 18 of 42 Korean universities and over 300,000 university students are studying in Seoul at any given time, and most of them choose not to leave the city after they graduate.

Seoul's tremendous importance in Korean life is not a recent phenomenon. Historically, Seoul has been the centre of political, educational and economic opportunity since it was established by the Yi Dynasty in 1392. Until 1910, the Yi Dynasty ruled Korea from this city. The name is derived from the ancient Korean word *sorabol* meaning the centre of everything. Despite its age, Seoul is relatively new compared to Korea's 5,000-year history. Before becoming the capital, the site now called Seoul was a small village known as Hanyang.

In assessing the economic progress of the city of Seoul, one must consider that when the Korean War ended in 1953, nearly every modern building was destroyed. The old capital building, City Hall, Myongdong Cathedral and a few other buildings are the only modern day remnants of pre-war Seoul. Also remaining after the war were five of the gates to a ten-mile wall which surrounded the capital city in ancient times. A few remnants of the wall itself can be seen on hillsides in the Seoul area. The major gates to this wall are used by Seoul residents as landmarks. Among these gates are Kwanghwamun, Namdaemun and Tongdaemun. It is relatively easy to get one's bearings in the Seoul area by looking for the tower on top of centrally located Namsan

Mountain. Before Seoul's tremendous growth, this mountain was the southern boundary of the city. Today, much of Seoul's development is occurring south of the Han River. The centre of downtown Seoul is City Hall Plaza from which an expansive ten-lane road runs to the old capital building.

As a general rule, residents work north of the Han River and live to the south, creating a horrendous traffic problem during the rush hours. The majority of Seoul's inhabitants live in newly constructed apartment complexes which tend to be individually owned. These 5- to 15-storey structures run in rows for several miles parallel to the Han River and many more such apartments are now under construction. A huge redevelopment project was recently completed which rejuvenated the Han River, lining the banks with a series of parks. Efforts are being made to reduce Seoul's considerable pollution levels. Whether this effort will be a success remains to be seen.

Nearly 90% of all Korean corporations have their headquarters in Seoul even though production facilities may be located much further south. Most government offices are also located in Seoul, although programmes have been implemented to move offices to satellite cities to ease congestion. Many government agencies (particularly those that are politically unpopular) have been moved to Kwachon City in Kyonggi Province, 40 minutes from downtown Seoul (south of the Han River). Seoul is where the power and culture of Korea resides so it attracts the "best and brightest." A few statistics illustrate this point: nearly 70% of the nation's currency circulates in the Seoul area; over 50% of Korea's passenger cars are registered in the city; and over 80 institutions of higher learning are located in Seoul, including all the most prestigious universities. A Korean proverb says, "Send your son to Seoul and your horse to Cheju." As a result, having an office in Seoul is a must for nearly all foreign companies.

Seoul is also the cultural centre of Korea. Both the National Museum of Korea and the National Folklore Museum are located near Kyongbokkung Palace. A large number of other fine museums can be found in Seoul. The Sejong Cultural Centre, the nation's major performance hall, is located in downtown Seoul. The National Theater is located in central Seoul on Namsan Mountain. Seoul is also the location of Kyongbokkung Palace, Changdokkung Palace, Toksugung Palace, and Changgyonggung Palace.

Special cities, provinces and industrial estates

Provinces and special cities

South Korea exercises sovereignty over nine of 14 Korean provinces (five provinces are in north Korea). These provinces and their population as of November 1 1986 are:

Kyonggi Province	5,075,449
Kangwon Province	1,750,707
North Chungchong Province	1,396,160
South Chungchang Province	3,009,102
North Cholla Province	2,192,133
South Cholla Province	3,848,897
North Kyongsang Province	3,035,391
South Kyongsang Province	3,516,829
Cheju Province	489,464

Each province is subdivided into many *kun* (equivalent to a county). For administrative purposes, Korea also has five *special cities* which have an administrative status equivalent to a province. These cities and their population as of November 1 1986 are:

Seoul	9,798,542
Pusan	3,578,844
Taegu	2,092,989
Inchon	1,441,131
Kwangju	928,851

Pusan is the second largest city in Korea and has historically been Korea's major port. The city has an excellent natural harbour and modern port facilities. Much of the nation's industrial plant is located near Pusan, particularly textile and footwear factories. Pusan is the terminus of the main rail line from Seoul as well as being close to Japan and the sea lanes to North America. Over half of Korea's total export trade passes through the port of Pusan.

Industrial estates and free export zones

To keep pace with the country's rapid economic development, the Korean government is actively promoting the establishment of *industrial estates*. Industries located in an industrial estate enjoy special privileges such as low land costs, good power and water supplies, high quality road networks, and various supporting facilities including special administrative support. Most important, however, are the tax benefits granted to foreign-invested enterprises which qualify under the Foreign Capital Inducement Law and which establish operations in an industrial estate. Bonded warehouses and factories may be established at any point in the industrial estates with the approval of the Office of Customs Administration.

Foreign investors may lease land in Korea. However, foreign nationals are subject to the Alien Land Acquisition Law which requires that foreign-invested firms in which a foreigner or foreigners hold over 50% of the equity obtain advance approval for land ownership in Korea.

The Korean government has also established *free export zones* to encourage direct foreign investment and exports. The first free export zone was established in Masan in 1970.

The Masan free export zone is located about 60 kilometres west of Ulsan. More than 150 companies have established operations at the Masan facility. In 1973, another free export zone was established near Iri which is 23 kilometres from the west coast port of Kunsan. These free export zones are special administrative tax-free areas with the characteristics of bonded areas where the application of various laws and regulations have been waived or relaxed entirely or in part.

Enterprises established in a free export zone should be in one of the following categories:

– Joint ventures between foreign and Korean investors or entirely foreign invested ventures

– Enterprise of a highly technical and labour-intensive nature with definite prospects for exporting.

– Located near other foreign companies.

The conveniences and advantages of these free export zones are as follows:

– Administrative procedures concerning foreign investment, joint ventures, approval of occupancy, and plant construction are simplified by the Office of Industrial Estates Administration

– Various necessary support and service facilities such as transportation, stevedoring, packing, repair and maintenance of machines and tools, and other common facilities needed by occupants are provided by the government.

Businesses locating in a free export zone can establish manufacturing facilities in government-owned buildings or in their own plants. The zones also offer housing facilities, medical facilities and other support services.

For political and national defence reasons, as well as to decrease population pressure on the Seoul area, the government has taken steps to locate industrial sites in southern Korea and in lesser developed provincial and coastal areas. Industrial estates have been established near the following Korean cities: Taegu, Pohang, Masan, Ulsan, Yosu, Iri, Kunsan, Taechon, Kwangyang, Mokpo, Kwangju, Kumi, Changwon, Chunchon and Wonju. The first industrial estate was established at Ulsan where nearly 20 large manufacturing and refining facilities have been constructed since it was established. The Hyundai business group dominates industrial activities in Ulsan. While other estates are not typically dominated by a single business group, most estates are dominated by a particular industry. A huge petrochemical complex has been constructed in Ulsan and another in Yochon. The Pohang Iron and Steel Company not surprisingly is located on the Pohang Industrial Estate. The bulk of foreign invested electronics companies have chosen to utilize the Kumi Industrial Estate. The machinery industry has chosen Changwon, a particularly large industrial estate located near Masan on Korea's southeast coast.

Regional rivalries

The homogeneity of the Korean people is striking in nearly all respects. Koreans have an enormous loyalty to their country and their culture. When allied against foreigners they form a cohesive front. However, on an intra-Korean level, rivalries between provinces and cities in Korea have for centuries created considerable friction. The principal conflict is between the two Cholla provinces and the two Kyongsang provinces. This rivalry can be traced to the Three Kingdoms period in Korea's history when the Shilla Dynasty ruled what is now Kyongsang and the Paekche Dynasty ruled what is now Cholla. After a protracted struggle, Shilla was eventually victorious. Cholla has since that time been known for its resistance to the authority of the central government. The rivalry between Kyongsang and Cholla continues today despite the pre-eminence of Seoul (which is located in the north, away from either province). This rivalry continues because Seoul's citizens continue to trace their origin to the province of their ancestors. The dialects spoken in the two provinces are very distinct, so it is relatively easy to trace an individual's home province. For this reason, many Koreans living in the capital city who are from Cholla have adopted the Seoul dialect. However, in their hearts, they continue to be loyal to the Cholla provinces.

The economic dominance of the Kyongsang Region was accelerated by the regime of President Park Chun Hee, a native son of Kyongsang. President Park openly pushed economic development in Kyongsang's cities (for example, example, Ulsan, Pohang and Changwon), as well as the restoration of the Shilla Dynasty's cultural treasures. The power of Kyongsang has also been magnified by the fact that the Korean Military Academy is located in Taegu, the capital of North Kyongsang Province. Both President Chun Do Hwan and President Roh Tae Woo (as well as many top military, government and business leaders), were born in Kyongsang and are graduates of the prestigious Kyongbok High School in Taegu.

This regional rivalry reached its apex when a nine-day uprising for free elections occurred in 1980 in Kwangju, a major city in South Cholla. President Chun put down the revolt with military force resulting in the death of at least 198 Koreans. Many more residents of Kwangju were wounded in this incident. The Kwangju incident is still a powerful political motivator for the people of Cholla. The role of the US Army in this incident (or its lack of a role) is a significant contributor to anti-American sentiment in Korea.

The provincial rivalry has also managed to find its way into current day Korean politics since Roh Tae Woo is from North Kyongsang, Kim Dae Jung is from South Cholla and Kim Young Sam is from South Kyongsang. It is no surprise that Roh Tae Woo is a conservative, Kim Young Sam a moderate leftist and Kim Dae Jung a radical leftist. Much of the vote in the 1987 Korean presidential election split along provincial lines. Kim Dae Jung and Kim Young Sam have been unable to unite to form a single opposition party in large part due to the regional rivalry.

North Korea

The Republic of Korea has been referred to as "Korea" throughout this book (unless otherwise indicated) despite the fact that it exercises authority only over the portion of the Korean peninsula south of the demilitarized zone established in 1953. It should be emphasized that the Republic of Korea claims sovereignty over the entire Korean peninsula. As a result, when a need arises to differentiate between the territory controlled by the southern democratic government and the territory controlled by the northern Democratic People's Republic of Korea, the words "south" and "north" are inserted before Korea. Note that north and south are not capitalized so as to emphasize that there is only one Korea.

The territory controlled by north Korea occupies 55% of the Korean peninsula. While accurate information is unavailable, the population of north Korea is believed to be approximately 21 million. Estimates place per capita GNP in north Korea at approximately US$1,100 and total GNP at US$23 billion. The economy of north Korea is socialized and centrally planned to an extent rivaling Stalinist USSR. Nearly all industrial production takes place in state-owned and operated factories. All agriculture in north Korea was collectivized by 1958. The focus of economic management has been on the development of heavy industry until very recently. This emphasis is in large part due to the need of the north Korean military for capital goods. These economic development efforts were funded by aid suppled by China, the USSR and other communist bloc nations. It has been estimated that the equivalent of US$1.5 billion in aid was given to north Korea by communist bloc nations from 1946 to 1960. Foreign aid granted to north Korea since that time has been substantially reduced. Most foreign currency is generated by selling coal and other resources as well as arms and military equipment. North Korea uses this military equipment to support terrorists throughout the world.

Recently, attempts at increased economic development by north Korea funded by borrowed capital have been a dismal failure, resulting in a complete cessation of principal and interest payments on US$77 million in loans to western banks (US banks are prohibited from making such loans).

In addition to the amounts owed to western banks, Japanese banks are also owed the equivalent of US$2 billion. Payment of these loans is not likely in the short-or medium-term. Despite rhetoric from north Korea that joint ventures with foreign companies would be welcomed, it is unlikely that any such ventures will be established. Such ventures are entirely at odds with the nature of the economy and the nation's economic and political philosophy.

The Democratic People's Republic of Korea has been controlled since its inception by the General Secretary, Kim Il Sung, a 75-year-old megalomaniac who was installed in power by Joseph Stalin in 1948. After becoming General Secretary, Kim Il Sung initially had to share some of his power with other pro-Soviet and pro-Chinese officials in the communist party. This power sharing lasted only a short time as Kim Il Sung began to consolidate his power in 1952 through a series of purges. These purges continue even today as Kim Il Sung is currently manoeuvering to install his son, Kim Jong Il, as his successor. If Kim Il Sung is successful, he will have created the communist bloc's first dynastic succession. Some opposition to this succession is thought to exist among the north Korean Central Committee and the military. For this reason, uncertainty and an inevitable power struggle following the death of north Korea's current dictator is expected. O Jin U, the north Korean Defence Minister, is reported to be in Moscow recuperating from an automobile accident. In some instances, "accidents" (sometimes fatal) happen to political figures who fall out of favor. O Jin U's accident appears to have been genuine.

General Secretary Kim Il Sung has created a personality cult which can be matched by the leader of no other nation in the world. His official titles include "Great Leader," "Immortal Father" and "Genius of the Revolution." Portraits and statues of the "Great Leader" are everywhere. Attempts have recently been made to glorify Kim Jong Il in similar ways as part of the preparation for his succession.

Kim Il Sung has installed a political and economic system he refers to as *juche* (which can be roughly translated as self-reliance). The juche system combines the worst of communism and isolationism and for that reason has been a dismal failure. The Great Leader's economic policies have produced the lowest standard of living in northeast Asia (outside of China) despite the imposition of forced labour on its citizens and abundant natural resources. After some initial success in the 1950s and 1960s, the economy of north Korea has become stagnant. The juche system has been used by the General Secretary as a tool to expel foreign-oriented dissidents from the communist party. Even communist party officials who are pro-Soviet and pro-Chinese have been purged on this pretence.

Defectors from north Korea describe an incredibly bleak existence. Men work at menial jobs for long hours with no possibility of advancement unless they are top party officials. Women also perform menial jobs, but are expected to cook and clean for the family as well. Children are immediately placed in nurseries and mothers are given access to infants during the day only because cow's milk is not available. Only the children of top communist party officials can expect to be educated beyond high school. Party officials enjoy a cultured and pampered life in marked contrast to ordinary workers.

Some visitors to south Korea are concerned about the danger of an attack by communist north Korea. This is particularly true for visitors who realize that Seoul is only approximately 30 miles from the demilitarized zone marking the dividing line between the two halves of Korea. A sightseeing trip to the truce village at Panmunjom in the demilitarized zone is a sobering experience. The massive military build-up and defence fortifications established by south Korea and its American allies to stop an attack by north Korea are clearly evident. The stony stares of the north Korean soldiers assigned to the truce village at Panmunjom and their Soviet-style uniforms are frightening.

The south Korean government holds a monthly civil defence drill to remind citizens of the ever-present danger. One day a month (around the 15th calendar day), an air raid siren will warn all citizens and visitors to take shelter. All cars and other vehicles must immediately pull to the side of the road. A legion of air raid marshalls and police will quickly appear to ensure full compliance. Occasionally, tanks or trucks full of soldiers will speed by an onlooker during the air raid drill which lasts approximately 20 minutes.

In the opinion of most experts, the danger from a north Korean military attack is minimal. Neither the People's Republic of China nor the USSR (both substantial benefactors of north Korea), has any interest in a regional conflict over the Korean peninsula. Although General Secretary Kim Il Sung is not entirely rational or stable, it is readily apparent to senior generals in the north Korean armed forces that an attack of south Korea would be a disaster. The combination of south Korea's 600,000 strong armed forces, the National Police Force of 120,000 men, the homeland reserves of 3.4 million men, and the support of the 41,000 member Eighth US Army (as well as air and naval support from US forces based in Japan), would result in certain defeat for north Korea. Only the direct involvement of the People's Republic of China or the USSR would prevent a total victory by south Korea.

Furthermore, the US military forces in Korea possess tactical nuclear weapons which could be used in defence of south Korea. The use of such weapons in north Korean territory to cut off supply lines would be likely if the safety of US forces and the Republic of Korea was at stake. Some knowledgeable observers have said that US military forces originally sent to protect the south from the north are now keeping the south from marching north to reunify the country. As south Korea's economy grows further, the chance of an attack by north Korea will diminish proportionately.

Currently, north Korea's GNP is less than one-fifth that of the south. With south Korea's economy growing at 12% and more per year, north Korea may need to devote approximately 40% of its GNP in years to come to the military. Attaining this level of expenditure is simply not possible for north Korea.

Currently, Korean and US forces have been formed into a Combined Forces Command, the only one of its kind in the world. Theoretically, a US four-star army general is commander-in-chief of this combined army. During the election campaign, President Roh stated that by the early 1990s south Korea would be able to defend itself without assistance. In fact, if current expenditure levels remain constant, south Korea will soon become a major military power in Asia. This may allow a partial military withdrawal in the mid-1990s.

One danger (particularly during the 1988 Seoul Summer Olympic Games) is the possibility of isolated terrorist activity sponsored by or performed directly by north Korea, such as the airport bombing which occurred during the 1986 Asian Games. This terrorist attack at Kimpo Airport near Seoul resulted in the death of several innocent Koreans and was intended to disrupt the Asian Games. For this reason, security in Korea at airports and other public places will always be extremely tight. Passengers and baggage are carefully searched on any departing flight.

Competitors and officials from a record 161 countries plan to attend the 1988 Summer Olympic Games, including nearly all communist nations. For this reason, an attack or major incident when these official delegations will be in Seoul is very unlikely.

People

People are Korea's major resource, especially since the country has few natural resources. At the end of December 31 1987, Korea's population was 42.34 million. With the exception of small city-states (such as Monaco), Korea has the second highest population density in the world. The total labour force in Korea has increased more rapidly than population due to increases in the number of working age individuals and increased female participation.

Korea's labour force is extremely well-educated by developing country standards. Over 97% of the population is literate, a significant accomplishment given the large number of Chinese characters one must know to read in Korea. Compulsory primary education and very high rates of middle school, high school and university graduation (fueled by Confucian emphasis on education) has created high education levels. Korea has a very large number of colleges and universities. In fact, due to a shortage of upper level positions, there is a surplus of university graduates in Korea, leading to significant *underemployment* (people working in jobs for which they are over-qualified). Economic

Planning Board statistics reveal that during 1987, 60% of the unemployed in Seoul were high school and college graduates. Many Korean students seek postgraduate education abroad for this reason. It has been reported that over 40,000 Koreans were in the US for educational purposes in 1987. It has also been reported that the Korean government employs more PhDs in economics granted by US universities than the US government does.

The number of workers is expected to rise by approximately 2% in 1988. The economically active population is expected to grow at a rate of 1.95% from 1986-91 and then fall to 1.64% from 1992-96, as the post-Korean War *baby boomers* will have already joined the work force. The 1985 Economic Planning Board Census (which is conducted every five years) revealed that in that year Korea's economically active population was approximately 15 million or 54.6% of the total population of Korea. Over 60% of these workers were male. Unemployment was 3.1% in 1987 (0.7% better than 1986), but is expected to rise to 3.7% in 1988 despite the creation of 354,000 new jobs by the expanding economy. The Economic Planning Board has reported that in November 1987, Seoul had an unemployment rate of 5.9%. Much of this unemployment is caused by Koreans graduating from universities located in Seoul who do not want to relocate.

In 1986, the US Bureau of Labour Statistics estimated the hourly cost for production workers in Korea to be US$1.38 per hour. A less accurate but more recent study placed the hourly wage rate in Korea at US$1.57 per hour, lower than Singapore (US$2.40), Taiwan (US$2.27) and Hong Kong (US$1.89). Wages in Korea have increased substantially since both of these studies were made. Labour turnover averages around 5% annually. Women workers leave their employers at a higher rate, particularly in assembly and manufacturing industries.

Union activity is expected to increase in the future in Korea. Knowledgeable observers have estimated that union and union-eligible Korean employees number around eight million out of a total work force of 20 million. It has been traditional for organized labour to press for wage and benefit increases during a *spring labour offensive*. As part of the spring offensive, Korean unions demanded a 30-35% pay raise for 1988. The employers countered with an offer between 8-15%. The Korean Employers Association predicts a wage rise of 10-14% during 1988.

Real wage increases of approximately 15-20% per year for Korean workers are expected over the next three to five years. In the past, clerical workers have had higher raises than factory workers despite their relatively higher wage scale. In 1987, clerical workers' wages rose by an average of 19.8% while increases in assembly workers' wages averaged only 10.1%. Increases should be higher for clerical workers than for assembly workers in 1988 if government policymakers have their way. Wage increases will affect both foreign-owned and domestic companies.

The labour disputes which occurred in 1987 were unprecedented in the history of Korea. More than 3,400 labour disputes were reported from June to September 1987. This is over 20 times the number of reported disputes in 1986. The Federation of Korean Trade Unions released figures showing an increase in its membership after the 1987 labour disputes from 1.1 million to 1.5 million. The Ministry of Labour reported a 40% rise in the number of unions over the same period. In addition to reported labour disputes, there were many thousands of unreported labour disputes. Korea's labour laws require labour disputes to be reported to be legal. During the labour turmoil during 1987, normal reporting procedures were not followed as the government was overwhelmed by the pressure from labour groups and the middle class.

While labour groups have pushed for both increased compensation and increased power to organize and control the work place, worker gains in compensation and benefits (shorter working hours, increased educational subsidies for children, health care) have been more significant than gains in worker autonomy and power. Korean businessmen, consistent with their authoritarian and paternalistic tradition, are simply unwilling to share power with labour to any significant degree. For this reason, the rise in wages and benefits during 1987 was much larger than might have been expected. It bears repeating that power sharing is simply not consistent with Korean character and values. The struggles for power between workers and management can (and often do) make labour negotiations very political in nature relative to their western counterparts.

Korean labour laws and requirements are often controlled more by practice than by laws, partly because laws and regulations change very rapidly. However, the following general comments should be useful. The Korean Constitution guarantees the rights of Korean workers to form unions and engage in collective bargaining. Labour practices in Korea are governed by the Labour Union Law. This law governs the rights of workers to conduct collective bargaining, defines what constitutes an unfair labour practice and creates the right to file a report of a labour dispute. Several other laws have been enacted to supplement the Labour Union Law, including the Labour Standards Law, the Labour Management Council Act, the Industrial Accident Compensation Law, the Industrial Safety and Health Law and the Labour Dispute Adjustment Act. In general, Korean labour laws allow the government to take a *parental* role in the labour dispute resolution process rather than creating a *level playing ground* for labour-management disputes. Korean labour laws also generally assume that there is no divergence of interest between workers and management. The emphasis of the laws is on preventing disputes and little is said about resolving disputes. This can create real problems once a dispute arises.

Foreign companies are held to a higher standard with regard to labour practices. As a result, extra care must be exercised by foreign companies in taking actions which may be deemed to be an unfair business practice.

The Korean government has recently adopted a minimum wage law which provides for a minimum wage of W111,000 per month in 13 designated manufacturing industries, and W117,000 in 17 specified manufacturing industries. A separate law requires that certain employers make a contribution of 1.5% of a worker's income to the National Pension Fund. Employees must also contribute 1.5% of their pay to the National Pension Fund. To assess the impact of the minimum wage law, it is useful to consider that the Economic Planning Board has reported that the average monthly income of urban households in urban areas in Korea was W562,060 during 1987. This was nearly a 17% increase over 1986.

The Labour Standards Law, enacted to protect Korean workers, applies to companies which employ more than a specified number of employees. The law requires that a labour contract be executed to control working hours, wages and working conditions with statutory restrictions on a company's ability to terminate, suspend, transfer or discipline an employee, requiring in most cases, *justifiable reason*. Should an employee be terminated, severance pay is calculated under a predetermined formula. Employers subject to the law must provide one day of paid holiday per month and additional paid holidays should certain attendance criteria be met. Sick leave standards are also specified in the law. Other benefits include condolence and funeral payments and certain allowances.

There is a relatively wide variation of wages between industries in Korea for skilled and unskilled workers. The lowest wages have generally been paid to workers in the textile, footwear, electrical machinery and wood products industries. Wages paid to these workers can be as little as W130,000 per month. Most of these workers are young single women living in company-owned dormitories. These women will typically work only until they are married. There are wide disparities between wages paid to women and men and between blue and white collar workers. Wages paid to workers in rural areas have traditionally been low. There are also large wage disparities between the amounts paid by large companies (particularly the chaebol) and the wages paid by small and medium companies. Even lower wages are paid by the many tiny independent subcontractors which supply parts to larger companies.

A large proportion of the salary paid to Korean workers is paid in the form of benefits and bonuses rather than salary. Most companies pay an annual bonus which ranges from four to eight months' pay. Companies also provide a wide range of non-cash benefits including dormitories, education loans, free meals and transportation allowances. Companies which employ a certain number of workers must also provide 50% of the premium for medical insurance coverage (the employees are responsible for the remaining portion). Fringe benefits can in some cases exceed 50% of a Korean employee's base pay.

Wage earners in Korea work long hours. The standard working week in Korea is 48 hours which is typically extended in manufacturing industries by paying overtime. In fact, Korean manufacturing employees worked an average of 54.7 hours per week in 1987, which is the world's highest rate. Workers in all Korean industries worked an average of 52.2 hours per week in 1987. Work weeks of the legal maximum of 60 hours are common (48 hours of regular time plus 12 hours of overtime) particularly in the manufacturing and assembly sector.

In the past, the Korean government actively discouraged, and in some cases suppressed, unions and labour organizing activity. But unions which recognize an identity of interest between labour and management have had some encouragement. Recent political developments in Korea have led the government to become more tolerant of labour activity. During the summer of 1987, the government generally refrained from cracking down on illegal strikes and attempted to encourage management and labour to deal directly with each other, hoping to avoid being criticized by both.

The most significant problem with disputes between management and labour in Korea is that neither group has any real experience in resolving their differences. In western countries, both parties have developed a form of etiquette which allows them to settle their differences. This etiquette has not yet evolved in Korea, nor are there individuals or groups available to act as mediators. For these reasons, labour disputes may take longer to resolve in Korea than expected. Further complicating the resolution of labour/management disputes is the Confucian abhorrence of open conflict.

In general, western companies treat their employees more favourably than Korean companies, partly because labour conditions are often better in the relevant foreign country and partly because Korean labour laws are more likely to be enforced on a foreign company. Care should be taken when selecting employees since it may be difficult or impossible to dismiss them. Furthermore, the Korean government has adopted a Veterans' Administration Law which requires that businesses of a certain size (measured in terms of numbers of employees) hire a certain percentage of veterans and their relatives. This law applies both to Korean and foreign companies. In order to avoid receiving a government directive to hire a particular individual (one who may often be unqualified for any position within the company), foreign companies should be careful to hire sufficient numbers of qualified individuals under the Veterans' Administration Law. For example, in 1986 an American manufacturing company received a directive to hire a recent college graduate with a French literature degree who had very marginal English language skills.

Chapter III
The role and structure of government

The role of government

Korean government officials consider themselves to be *directors of society* rather than *servants of the people*. The attitude adopted by the Korean government is extremely paternalistic due to their long-standing Confucian ideals. The western capitalist notion of economic development occurring as a result of the *invisible hand* of free competition is not widely accepted in Korea. Instead, the economic success of Korea is largely attributed to the firm guiding hand of the government.

Because of the acceptance of the government's control over business, Korean companies will nearly always respond to government directions even though they may not be legally binding. This practice of controlling companies through administrative guidance is widespread. This guidance usually takes the form of suggestions or warnings which have no binding force. Failing to comply with administrative guidance on the ground that it is not legally binding may result in disadvantageous treatment in future transactions for which government approval is required.

The role of the Korean government as directors of society is related to Korea's feudal past in two respects. First, the strong role of government in directing private affairs is in accord with traditional values. Korean kings ruled the nation with complete and unquestioned authority. Second, this strong role of government in society was necessary to move Korea from a feudal economy to a more modern economy. The unanswered question is whether the government will be capable of relinquishing its power as the economy matures and is less susceptible to management by administrative directive. In this regard, it is common for high level government officials to announce measures directed toward liberalization, deregulation and decontrol only to see their efforts thwarted by less enlightened junior officials in the bureaucracy. In general, the bureaucracy has no desire to see its power diminished. The involvement of Korean government officials in business matters in Korea is pervasive, ongoing and active. This is particularly true for international business transactions.

The relevant levels in a Korean government ministry are generally as follows:

Minister
Deputy Minister
Assistant Deputy Minister
Director General
Chief of Department
Deputy Department Chief
Section Chief
Deputy Section Chief
Staff

Government officials from the Chief of Department and below are commonly known as *working level* employees.

Structure of government

The highest level in the Korean government is the President. The president of Korea and his staff are commonly collectively referred to as the *Blue House*. The president of Korea has significantly more power than his counterparts in most foreign countries. He is the head of state, commander-in-chief of the military, chief diplomat and foreign policy maker. His power is increased due to the fact that legislation passed by the National Assembly always leaves much of the heart of the law to resolution by the issuance of a Presidential Decree. The ability to issue these decrees as well as to appoint or remove ministers and other government officials gives the Blue House immense power. The power of the president exists at local as well as national levels since even minor local officials such as governors and mayors are appointed by him.

The president is the chairman of the cabinet (sometimes called the *State Council*) which consists of the prime minister, the deputy prime minister, the heads of the government ministries and ministers of state. The president has authority to select who shall serve on the cabinet. The cabinet's purpose is to assist and advise the president in formulating and executing government policy. The Blue House has several agencies which are under the president's direct control. These agencies are the Presidential Secretariat, the Advisory Council on State Affairs, the Advisory Council on Peaceful Unification Policy, the National Security Council, the Board of Audit and Inspection and the Agency for National Security Planning (formerly the KCIA). The Presidential Secretariat, which is comprised of the secretary general and seven senior secretaries, is a very important part of the executive branch of government. The prime minister, assisted by presidential secretaries and the Office of Planning and Coordination, is responsible for coordinating and monitoring policy formulation and implementation.

The Korean National Assembly is a unicameral body composed of 299 members. Both regular and special legislative sessions are held by this legislative body. The National Assembly enacts laws, approves treaties and approves the national budget. The National Assembly must also approve the selection of the chief justice of the Supreme Court. The Korean Constitution gives the National Assembly some power to investigate and inspect government operations, but in general, serves as a forum for discussions of domestic and foreign policy. Most legislation is drafted by the executive branch which then submits the bill to the National Assembly for approval. Members of the National Assembly nearly always vote along party lines.

Judicial power in Korea is vested in a constitutionally empowered court system. The Korean judiciary is composed of four levels, the Supreme Court, appellate courts, district courts and the family courts. The chief justice of the Supreme Court is appointed by the president with the consent of the National Assembly. Other Supreme Court judges are appointed by the president. The chief justice appoints all other judges. Judges have jurisdiction over civil, criminal and administrative matters. Korea has adopted a civil law system which does not utilize a jury. Korean judges do not intervene in areas reserved for the legislative and executive branches of government in the manner common in the US.

The principal ministries and agencies

Economic Planning Board

The EPB is responsible for matters concerning the overall plans for the economic development of Korea, the formation and execution of the government budget and the coordination of resource mobilization plans. The minister of the EPB is concurrently the deputy prime minister and coordinates between ministries in matters related to economy and finance. The EPB has bureaus of Economic Planning, Budget, Fair Trade Policy, Project Evaluation, Economic Research, Statistics, Economic Coordination, and an Office of Planning and Management. The Offices of Supply and Fair Trade are also under the control of the EPB.

Ministry of Finance

The MOF is responsible for the Korean government's financial affairs, including finance, bills, currency, national bonds, accounts, taxation, customs, foreign exchange, external economic cooperation and the control of state-owned property and entities. The MOF also controls the Office of Monopoly, the Office of National Tax Administration and the Office of Customs Administration. The MOF has Bureaus of Treasury, Finance, Securities and Insurance, International Finance, Tax Systems, and Customs, and an Office of National Tax Judgment. The MOF also directs and supervises those private banks in which the government is the largest stockholder and the special banks established by the government.

Ministry of Trade and Industry

The MTI is responsible for commerce, foreign trade, the mining industry and standards for manufactured products. It is also responsible for trademarks, patents and cases of unfair competition. The MTI is also concerned with the promulgation of import and export regulations. In effect, it has jurisdiction over all manufacturing and trading in Korea. The MTI promulgates import-export regulations and is responsible for issuing export and import licences. The ministry has a number of important offices and bureaus including the Planning and Management Office, Trade Bureau, Basic Industry Bureau, Electronics Industry Bureau, International Trade Promotion Bureau, Textile and Consumer Goods Industry Bureau, Machinery Industry Bureau and Free Export Zone Administrative Office. The MTI also exercises great influence over a number of quasi-public organizations (for example, KOTRA and the KFTA).

Ministry of Justice

The MOJ is responsible for the prosecution of crimes, penal administration, control of exit from and entry into the country, protection of human rights, administration of civil and criminal justice and other legal affairs.

Ministry of Foreign Affairs

The MFA is responsible for matters concerning diplomacy, trade, treaties with foreign countries, other international agreements, and the protection and guidance of Korean nationals abroad. This ministry is also responsible for Korean missions to other countries and maintains relations with the diplomatic and consular representatives of foreign countries in Korea. The MFA is also responsible for certain matters concerning economic and financial relations with other countries.

Ministry of Science and Technology

The MOST is responsible for matters related to development and application of science and technology and management of industrial and technical manpower.

Ministry of Home Affairs

This Ministry is responsible for local administration, management of referenda, naturalization, civil emergency planning and the protection of the lives and property of citizens and national registries.

Ministry of Education

The MOE is responsible for the formulation and supervision of Korea's educational and scientific programmes.

Ministry of Sports

The MOS is responsible for national sports. The MOS controls the Korea Amateur Athletics Association and the Korean Olympic Committee.

Ministry of Agriculture and Fisheries

The MOAF is responsible for agriculture, fisheries and rural development.

Ministry of Construction

The MOC is responsible for establishing and coordinating plans for national land development, the control, conservation, utilization, development and renovation of land and water resources, and the construction of major infrastructure projects.

Ministry of Health and Social Services

The MOH is responsible for matters related to public health, epidemic prevention, public hygiene and sanitation, medical and pharmaceutical administration, public relief, emigration, family planning and various social welfare programmes.

Ministry of Transportation

The MOT controls matters concerning land transportation, marine transportation, air transportation and tourism. The MOT also supervises the Office of National Railroads and the Korea Maritime and Port Administration.

Ministry of Communications

This Ministry has responsibility for controlling postal affairs, telecommunications, postal exchanges, postal savings, postal pensions and national life insurance.

Ministry of Culture and Information

The MCI handles matters related to culture, arts, publications and broadcasting. Beginning in January 1989, the MCI will be split into a Ministry of Culture and a Ministry of Information.

Ministry of Labour

This Ministry is responsible for managing and coordinating matters related to the standardization of working conditions, occupation stability, job training, insurance and social welfare for workers and labour disputes.

Ministry of Energy and Resources

The MOER is responsible for all matters related to energy and other resources.

Ministry of National Defence

The MND is responsible for all defence related matters. The Army, Navy, Air Force and Homeland reserve forces are under the control of this ministry.

Office of Veterans Administration

This office is responsible for veterans' affairs matters, including loans, compensation, employment and insurance and other types of relief for disabled war veterans and policemen, bereaved families of soldiers, policemen and anti-Japanese independence fighters, students disabled in the student revolution of April 19 1960, and defectors from north Korea.

Office of Legislation

This office is responsible for drafting legislation, reviewing draft legislation, treaties, presidential decrees, ordinances, regulations and other related matters.

Dealing with government

In addition to the above ministries, there are numerous important offices, government corporations, quasi-government corporations, government research institutes and quasi-government associations. Among the more important of these organizations are the Bank of Korea, the Korea Development Institute and the Korea Foreign Traders Association. The names of these other important organizations are set out in the Business Glossary in the appendix.

Various government ministries will be involved in each business transaction depending upon its nature and scope. Often, more than one ministry will be involved and their conditions or requirements may even conflict. For example, several ministries have asserted jurisdiction over computer software (the Ministry of Trade and Industry and the Ministry of Communications). Another source of intra-government conflict is the requirement that a joint venture project be approved by both the Ministry of Finance and the *relevant ministry* which governs the specific industry. While the primary goal of the Ministry of Finance may be to encourage foreign capital and technology, the primary goal of the relevant ministry is to protect and foster the health of local industry. These conflicting goals can create conflicting policy. Thus, the position of a single official may not be indicative of Korean government policy. Furthermore, governmental authority is often delegated to other public and quasi-public organizations such as the Bank of Korea, foreign exchange banks and industry associations. In some cases, the head of a quasi-public organization can be more important than an official in a government ministry.

Foreign businessmen are often uncertain whether they should deal with the Korean government from the top-down or from the bottom-up. Which of these two approaches should be utilized in any particular case largely depends upon the nature of the matter. For issues which require a decision on a policy matter, a top-down approach will be most effective. Working level officials almost never have the discretion to make policy decisions without consulting with senior level officials. In nearly all other cases, the bottom-up approach is generally the most effective.

Working level government officials must be provided with a clear and comprehensive presentation of the matter in question. This presentation is very important since working level officials can make the approval considerably more difficult to obtain if they become alienated or hostile toward the foreign company. It should be remembered that because of the Korean government's ongoing role in business, working level officials will be dealt with on a continuing basis. The development of good working relationships can do much to smooth out problems with the Korean government.

Working level officials are generally much less accustomed to dealing with foreigners and are generally less sensitive to the needs of foreign companies than more senior officials. The foreign businessman should do everything possible to provide such officials with proper information. Comprehensive and wide ranging discussions should be held with the appropriate officials within the relevant ministry prior to submitting any application in order to determine their attitude toward the project and to help them understand the proposal. Any unusual aspects in the relevant business deal should be discussed carefully.

If working level government officials object to granting an approval, it will be difficult for a senior level official to overrule his objection due to the nature of the bureaucracy which exists in the ministry. Thus, it is a mistake to hold discussions only with senior level officials. In many cases, simultaneous discussions should be held at both senior and junior levels. Experienced foreign businessmen typically find this approach to be more effective since it is consistent with Korean attitudes concerning the social status of the parties involved in the discussion.

In dealing with government officials it is important to remember that the western concept of *rights* can have little meaning in Korea. Korean government officials believe that doing business in Korea is a *privilege* granted to a foreign company which brings with it a number of duties. Confucianism teaches that rights are unnecessary if duties are fulfilled properly. As a result of this belief, some government officials are accustomed to dealing with companies in a very paternalistic manner. They expect companies to follow government policies and directives without hearing claims that a *right* has been denied. The importance placed on rights in a society is evidenced by the number of lawyers. The US, which has a high rights consciousness, has approximately one lawyer for every 300 citizens. Korea has one lawyer for every 27,000 citizens. It has been reported that only 500 lawyers are actively engaged in private practice in Korea.

Korean government officials expect to receive requests from Korean businessmen in a non-aggressive, polite and non-argumentative manner. More importantly, the Korean businessman will strive at all costs to avoid damaging the pride of the relevant government official. Foreign businessmen should not adopt the Korean businessman's approach. If a foreign businessman fails to aggressively pursue his position by submitting his best arguments, his actions may convey a lack of interest or concern. However, overly harsh, direct and argumentative approaches are usually counterproductive even for foreigners. Presenting both the positive and negative aspects of the proposal is often a good idea, since government officials are accustomed to this type of presentation. One related matter to consider is that the type of aggressive Korean employee favoured by western companies may not be well received by government officials. In general, it is more important that working level officials be approached in the more traditional (i.e., less aggressive) manner since they are less familiar with western practices. These officials tend to be more nationalistic and less aware of macro-level foreign trade issues than their superiors.

The desire of a Korean company to enter into a business relationship is not sufficient to ensure Korean government approval of an agreement. The Korean government will nearly always independently review the fairness of the terms of the agreement and determine whether the transaction is in the national interest. As a result, the negotiation processes are not complete until final approval is received from the Korean government. The foreign businessman is often deceived by the fact that an agreement is executed prior to its submission to the government as a condition to approving the agreement. The foreign company should be prepared to make compromises as part of this government review process in order to obtain approval of the agreement. For that reason, it is prudent to anticipate changes which may be required. It is wise to negotiate the deal so that further changes or concessions would not make the transaction financially unattractive. Changes demanded by the government can go to the essence of the deal. For example, a royalty rate reduction or the elimination of an export restriction may be demanded by the government in return for granting an approval. It should be remembered that the Korean government's position will largely depend upon its view of the interests of the nation rather than the Korean company. The government approval process can effectively give the Korean company two bites at the apple if the foreign businessman is not careful.

While the foreign company should expect the Korean government to require changes in the agreement, the foreign company should not generally include provisions in the agreement which directly conflict with established administrative practice or published regulations. Once a government official has officially rejected an agreement or required a revision it is difficult to change his position. Informal and unofficial consultations with the government officials involved in the determination prior to the execution of the agreement can greatly facilitate and expedite the approval process. In effect, this process results in an informal pre-review of the agreement. Submission of the agreement itself together with the application then may become only a formality. During this informal pre-review process, unusual aspects of the deal can be explained in detail and justified. In this setting, government officials will not be embarrassed by a lack of knowledge on a specific matter under consideration. The pre-review process can also help avoid encountering unexpected conditions from the government during the approval process which may make the deal financially unattractive (for example a reduction in royalty rates or the elimination of restrictions on post-licence use of technology).

Legal system

Korea has adopted a civil law system rather than a common law system utilized in countries like the US and the UK. While concepts developed in Germany, France and other European countries have been used in formulating the civil law system, traditional Korean legal principles and attitudes have also been incorporated into the legal system. Korean statutes usually contain only general statements of the principles and considerations to be applied by the competent ministry in regulating the relevant transaction or conduct. This function is achieved both by specific statutory delegations of authority and by a willingness on the part of Korean government officials to creatively interpret the actual language of statutes. Korean government officials have a great amount of discretion and autonomy in interpreting laws and guidelines to the laws. This aspect of the Korean legal system emphasizes the importance of being familiar with recent decisions which have been made by the relevant ministry and having a good working relationship with government officials.

Government officials enjoy high status in keeping with Confucian teachings. Under a Confucian social structure, government officials form part of the upper class. Until recently, nearly every top student aspired to hold a position with the government. Such a job brings great prestige to the young man and more importantly, to his family.

In some situations, a Korean will take a position which is contrary to Korean law and will urge the foreign party to do the same. The foreign businessman must be extremely careful in agreeing to any such proposal. While a Korean might be excused in such a situation, the foreigner may not receive similar treatment. For example, foreign businessmen should not sign incorrect tax receipts or invoices when requested by a Korean despite statements such as "it's a formality" or "everyone does it." If a foreigner is caught engaging in illegal conduct the result can be disastrous both for him and his company.

The single most important rule concerning the necessity of obtaining advance approval from the Korean government is that any transaction involving a transfer of currency or resulting in a credit relationship between a resident of Korea and a non-resident must be approved by the Korean government. If foreign exchange approval is not obtained, no legal basis exists for conversion of Korean currency into foreign currency or for the remittance of funds from Korea. Retroactive approvals are not granted. This approval requirement exists in order for the Korean government to control the allocation of available foreign exchange and to stabilize the value of the domestic currency. As a practical matter, this foreign exchange approval requirement allows the Korean government to monitor every agreement involving a non-resident and a resident of Korea and to prevent any transaction considered to be contrary to the national interest of Korea.

Economic policy formulation

The president of Korea has final decision-making authority on economic issues. He is supported and advised in making these decisions by three organizations. The first organization is the Blue House staff which is headed by the senior presidential secretary for economic affairs. The second organization is the Economic Minister's Council which is chaired by the minister of the Economic Planning Board. The ministers of Trade and Industry, Finance, Agriculture, Energy, Resources, Telecommunications and Transportation are members of the Economic Minister's Council. The third adviser to the President on economic issues is the Economic Planning Board. To a lesser extent the Ministry of Foreign Affairs and the Ministry of Justice play a role in economic policy implementation. The Bank of Korea also plays a role in the economic policy formulation area, but its role is reduced because it is much less independent than in some foreign countries.

In terms of policy formulation, most observers believe that the Economic Planning Board plays the most significant role. This organization formulates the five-year economic development plans (EDPs), administers their implementation, prepares government budgets and coordinates government economic policy efforts. The Economic Planning Board is supported in its policy formulation function by the Korean Development Institute. This institute is a research and planning organization which provides technical support to the Economic Planning Board.

The Ministry of Trade and Industry has a very important a role in economic policy formulation. This ministry assumes the primary burden of collecting input from the other ministries and government organizations as well as receiving information from a wide range of industry and economic associations and for incorporating this data into the policy-making process.

Chapter IV
Foreign relations

The Ministry of Foreign Affairs (*MOFA*) is responsible for planning and implementing the nation's diplomatic and foreign relations policies. In 1986, this ministry had a staff in excess of 1000 and a budget of US$127 million. The MOFA is involved in negotiating nearly all trade and treaty issues as well as dealing with matters relating to Koreans residing abroad. The ministry is responsible for the 89 embassies, 34 consulates and two permanent missions which constitute its diplomatic presence abroad. All diplomatic staff, including ambassadors, consuls general and consuls are under its jurisdiction. Korea is a member of more than 60 international organizations and UN agencies as well as a huge number of government and private international organizations. The MOFA supervises these memberships and generally promotes Korea's diplomatic interests through the following bureaus:

>   Asian Affairs
>   American Affairs
>   European Affairs
>   African and Middle East Affairs
>   International Organizations
>   Treaties
>   Economic Research
>   Evaluation and Coordination
>   Overseas Residents Affairs
>   Telecommunications and Documents

The ministry also has offices of Planning and Management as well as Protocol.

The principal foreign policy task for Korea is achieving the peaceful reunification of the country. This task is so important that a cabinet-level National Unification Board has been created to work toward reunifying Korea. Each president of south Korea has made proposals to north Korea concerning national reunification, without any real success. The pattern of dialogue between the two Koreas appears to run in cycles, with periods of progress and reciprocal visits by political and cultural delegations being followed by disappointment (often caused by terrorist and military action by north Korea). North Korea's attempted assassination of President Chun in Burma, several cross-border incursions resulting in the death of Korean and US military personnel, the airport terrorist bombing during the 1986 Asian Games and the downing of a Korean Air commercial airliner over the Thai-Burma border by a time-controlled bomb have all served to dash the hopes of those who desire reunification.

Throughout much of its history, Korea's external political and economic relations were restricted to limited contacts with China and Japan. In general, relations with the Chinese were tributary in nature as Korea was effectively a loyal buffer state between China and Japan. Since most early contact with Japan involved invasions of the Korean peninsula by Japanese military forces, political ties between the two nations were not strong. A particularly severe and damaging occupation by Japanese troops under Hideyoshi is widely believed to be the cause of Korea's decision to bar contacts with the outside world as completely as possible and become a *hermit kingdom*.

Korea only began to emerge as an independent part of the world's foreign relations community after its liberation from Japanese colonial occupation in 1945. This development was immediately dealt a severe blow when the country was partitioned along the 38th parallel after negotiations to establish a unified country were a total failure. Because of the partitioning of the country and the economic devastation caused by the civil war, the south Korean government was initially required to direct its foreign relations efforts toward those countries which could provide economic and military assistance. For this reason, in the 1950s, south Korea concentrated on its relations with the US, UK, West Germany, France, and the Republic of China (Taiwan).

As Korea's economy and political situation became more stable, the government began a major effort to establish and foster diplomatic relations with the non-aligned nations of the world. The purpose of this effort was to upgrade the nation's status in the world community so that support for national reunification could be generated around the world.

Korea is now beginning a new stage in its diplomatic efforts by moving to increase the nation's ties to communist nations. This process is severely complicated by national partition, but recent events indicate that Korea has so much to offer on an economic level that it is inevitable that diplomatic relationships will be established.

Relations with the US

The political, economic and military ties between the US and south Korea continue to be very strong. The Reagan administration has been a strong supporter of both the Chun and Roh administrations. President Reagan visited Korea in the early part of his administration and received several visits from President Chun. The US State Department has repeatedly stated that the US supports the progress of democratization underway in south Korea. A complete withdrawal of US armed forces from south Korea is unlikely until at least the mid-1990s, particularly under a Republican administration, and even then the withdrawal would be phased over several years.

South Korea has never been colonized by the US or any other western power so the distrust of western military power which exists in many other Asian countries is not a significant factor. A reservoir of goodwill exists since many older Koreans remember the assistance which the US provided after the Korean War. Support for US foreign policy not directly affecting Korea is strong in Korea as a rule.

Anti-American sentiments exist primarily in younger Koreans and older Koreans who feel that they have not received their fair share of the fruits of economic development. A large majority of these individuals are supporters of Kim Dae Jung. Many Koreans who side with the opposition feel that the presence of US military forces has allowed the present military-affiliated Korean government to continue in power. Some of these same Koreans blame the US for not preventing the killings which occurred during the *Kwangju incident* in 1980 under President Chun.

The state of the bilateral trading relationship between Korea and the US is uncertain. The Roh administration seeks time to consolidate its grip on political power. The Reagan administration seeks improved trade statistics to defeat protectionist sentiment in the US and to increase political support. Achieving both Korean and American goals simultaneously will be very difficult. The complete graduation of Korea from GSP benefits was a strong signal from the US that the size of the current bilateral trade surplus in Korea's favour will not be tolerated. However, if President Roh makes too many trade and other concessions to the US his support from the domestic electorate will be damaged, as Koreans are extremely nationalistic and sensitive to external pressure. In the final analysis, Korea is so dependent on the US market for economic prosperity that it must move to reduce the trade surplus by increasing imports and diversifying exports. Whether Korea moves quickly enough to avoid sanctions from the US remains to be seen.

Relations with Japan

Korea has had a love-hate relationship with Japan since the end of the Second World War. Memories of atrocities committed during the colonial occupation by Japan are still vivid in the minds of most older Koreans. The ill feelings against Japan are so strong that diplomatic relations between the two countries were established in 1964 and only after intense domestic debate in Korea. The relationship between the two countries is complicated by the large number of long-term Korean residents of Japan who are a legacy of the colonial period. These Koreans are subject to severe discrimination in Japan and are not allowed Japanese citizenship. A number of these overseas Koreans are members of groups supported by north Korea and engage in a range of activities against the government and people of south Korea. These activities range from terrorism (such as the bomb which downed a Korean Air commercial airliner over the Thai-Burma border in 1987), to a wide range of propaganda. Also complicating matters is the significant amount of trade which is conducted between Japan and north Korea.

Japan is second only to the US in its economic importance to Korea. Korea's development is closely linked to the Japanese economy, since many parts and components and much of the technology used by Korea come from Japan. In the years to come, Japan will also become an increasingly important market for finished goods made in Korea.

Korea has patterned many of its government institutions and legislation on the Japanese model. While substantial differences in the political and economic systems of the two countries exist, Korea will continue to look to Japan for ideas on ways to encourage and manage growth as well as how to assimilate western technology into their society without losing their national identity.

From the Japanese perspective, there is some concern about the growing military might of Korea. Japan simply cannot keep pace with Korea militarily if Korea's GNP continues to grow at current rates and military expenditure remains at a level equal to 6% of GNP. Some commentators believe that Japanese politicians would prefer a divided Korea to reduce the threat of Korea to Japan.

Soon after taking office, President Roh initiated a series of diplomatic contacts with Japan. Roh can be expected to further diplomatic relations between the two countries to enhance his image as an international statesman. High on his agenda will be the treatment of Koreans residing in Japan, and the opening of Japanese markets to Korean manufactured goods.

Relations with the USSR

The USSR has been caught in a triangular relationship with north Korea and the People's Republic of China. North Korea has been successful for many years in playing one communist benefactor off against the other in an attempt to maximize foreign aid and diplomatic support. There is some evidence that the USSR has grown tired of this game. However, they will do nothing to encourage north Korea to ally itself more closely with China. In dealing with south Korea, the USSR will also be wary of provoking Japan to ally itself with China. As the USSR has a significant interest in maintaining the political and military status quo in the region, the increasing strength and sophistication of south Korea's military will be a major concern to the leadership for years to come.

The decision of the USSR to send athletes to the 1986 Asian Games and the 1988 Summer Olympic Games held in Seoul is a major development in political and economic relations between the two countries. The next step would be to increase the level of indirect and direct trade between the two nations. Despite this *sports diplomacy*, nearly all such trade is likely to be conducted by intermediaries in the immediate future. It should be noted that direct trade contacts between the USSR and Korean companies are increasing. The three major Korean electronics companies participated in a Leningrad electronics exhibition in May 1988 and Jindo, Korea's largest fur producer, has entered into a joint venture to produce US$20 million of fur garments per year near Moscow. Exports of high technology products to the USSR will need to be handled carefully by south Korea to avoid a *Toshiba-type* illegal technology sale which would upset the US. Trade between Korea and the USSR and Eastern Europe was estimated to have a value of US$200 to US$300 million in 1986. Significant growth (approaching 50% per year) in this trade is expected over the short- and medium-term.

One interesting sign of a thaw in Seoul-East Bloc relations was the announcement that Poland and Hungary will be opening a trade office in Seoul in late 1988. East Germany and Yugoslavia are rumoured to be considering the same move. All indications are that south Korea intends to reciprocate. The Korea Overseas Trade Promotion office established a representative office in Hungary in December 1987. In the private sector, the Ssangyong business group is reported to be planning to open liaison offices in Hungary and Yugoslavia in late 1988.

Relations with the People's Republic of China

President Roh has made it well-known that he desires a wider range of diplomatic, economic, sports and cultural ties between Korea and the People's Republic of China. It has even been reported in the press that he wishes to visit Peking during the first years of his administration. Certainly, the cultural and athletic exchanges in recent years have created better relations between the two countries. This process should be furthered when Chinese athletes visit south Korea to participate in the 1988 Summer Olympic Games.

Trade with the People's Republic of China has some significant problems to resolve before it can increase markedly. In November 1987, Hu Qili, a member of the Politburo Standing Committee, was quoted in a Japanese newspaper as stating "If north Korea agrees, direct trade can be pursued." It was also reported by the Yonhap Press Agency that Chinese officials would not accept south Korean documents should a south Korean vessel arrive at a Chinese port. For the present, trade between the two countries can be expected to proceed through Hong Kong, Japan and Singapore. The other major problem related to increasing the amount of trade is China's foreign currency shortage. For this reason, significant amounts of barter trade can be expected in the years to come.

As many as 50 Korean companies are believed to be currently doing business in the People's Republic of China. The most significant investment by a Korean company to date is a US$10 million refrigerator plant constructed by Daewoo in Fujian Province. Other Korean companies said to be involved in joint ventures with Chinese firms include Cheil Sugar, Nhong Shim, Samyang and Sammi Steel. Many of these ventures involve Hong Kong subsidiaries of the Korean company. Lucky-Goldstar is scheduled to exhibit products at a Shanghai trade fair in October 1988.

The principal Chinese exports to Korea are coal, oil, lumber, minerals and grains. It has been reported that Pohang Iron and Steel Co. has bought large quantities of Chinese coal. Principal Korean exports to China are industrial and consumer goods (such as televisions, steel and refrigerators).

To increase trade with the People's Republic of China, the Roh administration has announced an ambitious port and industrial facilities expansion effort along Korea's western coast. For political reasons, most facilities will be located in the economically less developed Cholla provinces. Korean companies have begun to gear up for this trade expansion effort by encouraging their employees to develop language skills. Sales of Chinese language books are said to be booming. Trade seminars on doing business in the People's Republic of China have proliferated in Seoul since 1987.

During 1986, trade between the People's Republic of China and Korea is believed to have consisted of US$600 million in exports and US$700 million in imports (giving Korea a US$100 million trade deficit with China). Total trade volume in 1987 was believed to be US$2 billion. Total trade in 1988 may grow by as much as 50%. These trade figures are only estimates, given the lack of diplomatic ties between the two countries. In addition, the indirect nature of much of the trade makes the calculation of bilateral trade levels particularly difficult. Clearly, however, the level of trade is increasing very rapidly.

In many respects the economies of the two countries are complementary. In the near- and medium-term, Korea will emphasize technology and capital-intensive industries while China concentrates on labour-intensive industries. Korea is resource-poor but has technology suitable for developing countries; China has vast resources and low-cost labour. The physical closeness of the two nations and China's desire for economic modernization is too strong to prevent trade from growing despite huge political differences. In the short-term, the Asian concept of *appearance differing from reality* is likely to prevail. The appearance will be shown by political distance while the reality will be expressed through increased levels of trade largely conducted through intermediaries.

President Roh Tae Woo has ushered in the era of the ordinary person (top); Kim Dae Jung is widely considered to be more radical than most opposition leaders and more of an idealist than a pragmatist (bottom left); and Kim Young Sam – leader of the largest opposition group in Korea (bottom right).

Chapter V
Domestic politics

Roh Tae Woo and the Democratic Justice Party

The victory of Roh Tae Woo, chairman of the ruling Democratic Justice Party (DJP), in the December 16 1987 election to a five-year term as president of Korea should bring about a degree of stability. His victory in the presidential election can be attributed less to his popularity (he won only 36.6% of the total vote) than to a divisive split in the opposition. Roh received approximately two million more votes than Kim Young Sam (the runner-up), but four million votes less than the opposition as a whole. Polls after the election reveal that Roh's support was strongest in the countryside and among less-educated voters. Despite allegations of fraud, President Roh's victory is widely believed to have been genuine and solely the result of opposition discord.

President Roh Tae Woo is 55 years old and a retired four-star general. Roh was born in Taegu into a family headed by a local government official. He is married and has two children. President Roh's rise to power can in large part be attributed to his longtime friendship with ex-President Chun Doo Hwan. Both Roh and Chun are graduates of the 11th Class of the Korean Military Academy. They also attended a training programme together in the US in 1959. Roh served as the commander of a Korean battalion in the Vietnam War and is believed to have played a key role in the 1979 coup that brought President Chun to power. Roh moved the troops under his command into Seoul when the army assumed control of the government after President Park Chung Hee was assassinated. He was also instrumental in assisting Chun in consolidating power after the coup and has held a number of key positions, including Minister of Political Affairs, Minister of Home Affairs, Minister of Sports, President of the Seoul Olympic Organizing Committee (SLOOC), and the titular head of the Democratic Justice Party immediately before the 1987 presidential election.

Backers of Roh have attempted to portray him as a moderate who can keep the powerful Korean military in check. Roh worked hard to distance himself from Chun during the presidential election. Roh's claim to this moderate position is largely based upon his surprising eight-point *Special Declaration for Grand Harmony and Progress Toward a Great Nation*. In his declaration, he proposed: (1) a direct presidential election; (2) fair elections; (3) amnesty for most political prisoners; (4) respect for human rights; (5) freedom of press; (6) autonomy for all social sectors, including college campuses; (7) sound political activities; and (8) social reform to promote a healthy society. Roh's announcement stopped violent mass protests which were then sweeping the country. Two days later President Chun endorsed Roh's plan. Some believe that Roh made the announcement without Chun's consent and Chun had no choice but to comply, given the overwhelming public support generated by the announcement. By October, a new constitution had been worked out with the opposition and then approved in a national plebiscite.

Since his election, Roh has shown that he is both a strong president and an ordinary private citizen. Roh has repeatedly vowed to usher in the "era of the ordinary person" by furthering the interests of this group over the power elite. At his inauguration Roh stated that "The day when freedoms and human rights could be slighted in the name of economic growth and national security has ended." He has appointed a moderate prime minister and spurned some of the *imperial* perquisites of the office of president. He reportedly refuses to be called *gak-ha* or *excellency* and carries his own briefcase. Helping President Roh is the fact that he has a more relaxed personality than his predecessor. These image-improving steps have had some degree of success since a KBS-Seoul National University poll showed that President Roh's popularity had nearly doubled after his first few weeks in office.

Undercutting Roh's credibility as a man untainted by previous regimes was his decision to retain several key officials from the government of President Chun. The cabinet announced by Roh on February 19 1988 was bound to upset both liberals and conservatives in the ruling DJP as well as to produce a storm of criticism from opposition politicians. The cabinet retains seven ministers in their current posts, including four key ministers. The cabinet line-up is as follows:

		Age
*Prime Minister	Lee Hyun Jae	59
*Deputy Prime Minister and EPB Minister	Rha Woong Bae	54
Foreign Affairs	Choi Kwang Soo	53
Home Affairs	Lee Sang Hee	56
Finance	Sakong Il	48
Justice	Chung Hae Chang	51
*National Defence	Oh Ja Bok	58
*Education	Kim Young Sik	58
Sports	Cho Sang Ho	62
*Agriculture, Forestry and Fisheries	Yun Kun Hwan	59
*Trade and Industry	Ahn Byong Wha	57

*Energy and Resources	Lee Bong Suh	52
Construction	Choi Dong Sub	53
*Health and Social Affairs	Kwon E Hyock	65
*Labour Affairs	Choi Myung Hun	59
*Transportation	Rhee Bomb June	60
Communications	Oh Myung	48
*Culture and Information	Chung Han Mo	65
*Government Administration	Kim Yong Kap	52
*Science and Technology	Lee Kwan	58
*National Unification	Lee Hong Koo	54
Presidential Secretary General	Hong Sung Chul	62
*Office of Legislation	Hyun Hong Choo	48
*Patriots and Veterans Affairs Agency	Juhn Suk Hong	54

(* newly designated ministers)

The ministers of Foreign Affairs, Home Affairs, Finance, Justice, Sports, Construction and Communications were retained from the Chun government.

The prime minister selected by Roh Tae Woo is Lee Hyun Jae, a 58-year-old economist and former president of Seoul National University. The Prime Minister is considered to be a moderate who has in the past exercised tolerance in dealing with student protests. He is a native of the Chungchong region, which is important since both Roh and DJP head Chae Mun Shick are from Kyongsang Province. Prior to his election he had never served as a political official.

Deputy Prime Minister Rha has for some time been an advocate of trade liberalization and economic decontrol by the government. He has worked in business, government and politics and thus has a good background to deal with broad policy matters. He is particularly interested in international trade regulations. Rha is a graduate of Seoul National University and obtained a graduate degree from the University of California in 1968. Rha and Park Seung (who was named to serve as senior presidential secretary for economic affairs) will compete for the president's ear on economic issues. Park Seung is a graduate of Seoul National University and has a graduate degree from New York University. Finance Minister Sakong Il also has a leading role in developing and implementing the government's economic policies. Controlling the speed of the devaluation of the won against foreign currencies and reducing inflation are his major priorities. Sakong Il is a native of Kyongsang and has a graduate degree from the University of Southern California, Los Angeles.

Another key figure in the government is the Presidential Secretary Hong Sung Chul. Hong was born in what today is north Korea. He is a leader of political groups tracing their lineage to north Korean territory and was instrumental in delivering their political support to Roh. He is a graduate of Seoul National University and a retired Korean Marine Corps Colonel.

One major surprise was the choice of Ahn Byong Wha to head the Ministry of Trade and Industry. Ahn was formerly the head of a number of private and government businesses including Pohang Iron and Steel Co. (POSCO) and one of only a few businessmen to achieve cabinet rank in Korea. Ahn's most recent position was as head of Korea Heavy Industries & Construction Co. Yun Kun Hwan, previously president of the National Agricultural Cooperatives Federation, is the new Minister of Agriculture, Forestry and Fisheries. The home minister is Lee Choon Koo.

In a county where political power sharing can never be expected to take place without a major struggle, many in the DJP would like to amend the Constitution to create a cabinet-style government headed by a prime minister to perpetuate their political power. The opposition will resist such a move at all costs.

The two Kims

The principal opposition leaders are known as *the two Kims*, Kim Young Sam and Kim Dae Jung. After losing the presidential election, the two Kims announced that they planned to merge both parties and form a new political organization to present a unified front for the National Assembly elections. This move is consistent with traditional Korean party politics, as political organizations are nearly always created to serve the interests of a principal politician and his followers rather than advance a slate of issues. What would have been unique about the merger would have been the formation of a political party in which power is shared by several major figures. To be effective, the two Kims would have had to bring other opposition figures back into the fold. Such a collegial power structure is completely at odds with Korean attitudes and values and is unlikely to take place. The likelihood of a rift between the two Kims is increased by their rivalry and by the differences between their political bases. Kim Young Sam is considered to be a centrist and Kim Dae Jung a leftist. An indication of the instability of the opposition as a whole is the fact that the proposed new party would have been the fourth in less than a year that the Kims have led either jointly or separately.

Prior to the National Assembly elections, Kim Young Sam resigned as the titular leader of the Reunification Democratic Party (RPD). However, nearly all observers believe that he remained firmly in control. After the election he was re-elected to the top leadership position. Kim Young Sam's party is the largest opposition group in Korea in terms of membership. The RPD's strength lies primarily in its support from the middle class. In general, its followers favour democratization and the demilitarization of government rather than significant social change. Because of the size of his party, Kim Young Sam has insisted that other opposition parties join forces with him. Kim Young Sam is 59 years old, was born in Pusan and has been a professional politician all of his adult life. He has a reputation of being an excellent political dealmaker, a pragmatist and is more moderate politically than many other principal opposition figures. Kim Young Sam appears to have accepted the result in the presidential election of 1987 to a greater degree than other opposition figures and has adopted a relatively conciliatory posture. Following the election he met with President Roh and publicly pledged to work with him to bring genuine democracy to Korea.

Kim Dae Jung is the second of the two Kims. This 63-year-old also resigned as leader of the Party for Peace and Democracy (PPD) before the 1987 election, but remained firmly in control and was also re-elected as leader. He is widely considered to be more radical politically than most opposition leaders and more of an idealist than a pragmatist. He is an excellent public speaker who has adopted a populist political strategy, professing to be the candidate of the common man and those who have not shared equally in Korea's economic development. Much of his political support is based in the economically disenfranchised Cholla Region. Over the course of his political career, Kim Dae Jung has been sentenced to death, allegedly taken from Tokyo by the Korean Central Intelligence Agency in 1973, jailed, and exiled. He was nearly elected president of Korea in a 1971 election against the then president, Park Chung Hee.

Both Kim Dae Jung and Kim Young Sam have been widely criticized for failing to agree on a single presidential candidate in the 1987 election, as together they received 55% of the vote. The possibility of the two Kims agreeing on a single leader is believed by most commentators to be remote. As a result of the inability of the two Kims to reach an agreement, a new group of political leaders of the opposition is likely to emerge. It is very possible that the leaders will come from a group of opposition National Assemblymen who broke away from the two Kims even before the elections. Any such leader is likely to form a new party rather than campaign under the banner of one of the existing opposition parties. Many observers blame the selfish and egotistical behaviour of the two Kims for the lack of a new generation of political leaders of the opposition.

Some political commentators have begun to refer to the *three Kims* by including Kim Jong Pil, a former army general and prime minister under the assassinated president, Park Chung Hee. At the time of Park's death, Kim Jong Pil would have been his logical successor. His New Korea Democratic Party (NKDP) is as conservative as the DJP and not a true opposition party, but another group angling for power. Economic and social policies would not be significantly changed if the DJP were to share power with the NKDP, only some of the people in power.

One common denominator the three Kims share is an abiding hatred for the Chun regime. Each feels that Chun and his colleagues prevented them from becoming president and will continue to campaign for the presidency during Roh's term of office. This should make them less obstructionist and more statesmanlike, qualities that Roh will be counting on as he tries to effectively govern.

Elections for the National Assembly were held on April 26 1988. Nearly 20 parties fielded a range of candidates for office in the election. Contrary to the predictions of nearly all analysts, the ruling DJP failed to win a majority of seats in the National Assembly. The elections were for 224 seats at the district level, and 75 additional seats to be filled on the basis of proportional representation. On this basis, the DJP will have 125 seats out of 299 in the National Assembly, well short of a majority. Kim Dae Jung's PPD won 70 seats and Kim Young Sam's RDP, 59, but the major surprise was Kim Jong Pil's NKDP which will have 35 seats. The remaining 10 seats go to independent candidates. Roh Tae Woo's party attracted less than 34% of the popular vote, lower than the level obtained in the presidential election. Kim Young Sam's party obtained approximately 24%, Kim Dae Jung's party took 19.2%, and Kim Jong Pil's party 15.4%.

The biggest loser in the National Assembly election was Roh Tae Woo, who now must contend with a hostile legislature. This is the first time a Korean president will be required to deal with such a situation, so he will be covering new ground. However, his power is such that it will not prevent him from governing effectively. The DJP made a major mistake in agreeing to a one candidate per district election since its candidates registered the second greatest number of votes in many districts.

The biggest winner in the National Assembly election was Kim Dae Jung, who translated a relatively low vote tally into a disproportionately large number of seats under the one candidate per district system. Kim Young Pil was also a big winner and will now have a national platform to address the public. The DJP and Roh Tae Woo will need Kim Jong Pil to effectively control the National Assembly. Forging a coalition with Kim Jong Pil will not be easy for the DJP, because of his hatred for the Chun regime.

A number of factors resulted in the poor showing by the DJP. The low voter turnout (only 75% of registered voters) was far less than the 89.2% who voted in the 1987 presidential election and was one of the major factors. Also hurting Roh Tae Woo's party were the scandals involving ex-President Chun and his family, the one representative per district election system, and a desire on the part of the electorate to check the president's power. Regional, as opposed to national, issues tended to be more important in deciding this election. The voters favouring the opposition also didn't waste votes on single-issue splinter candidates and instead backed potential winners from the opposition.

A major disappointment of the election was the furthering of the regionalization process. Kim Dae Jung's PPD captured all but one seat in the Cholla region, Kim Jong Pil's party dominated in Chunchong and Kim Young Sam took 14 of 15 seats in Pusan. President Roh's party dominated as usual in the Kyongsang region. As a result of the election, there is no truly national party. This is not a positive development for democracy of Korea. In some respects, current Korean politics can be described as feudal, a modern-day *four kingdoms* period.

It is certain that Roh Tae Woo and the DJP will need to grant some important assignments in the National Assembly to the opposition. While the key committee posts, such as foreign affairs, justice, defence, finance and home affairs will be retained by the DJP, the opposition can expect some real involvement in government. This will be a positive development since the opposition will need to participate in governing rather than simply opposing government policy. This will give opposition members much needed experience.

The role the National Assembly will play in the new government is unclear. In the past the legislature was little more than a forum for political posturing, *rubber-stamping* the policies and legislation proposed by the executive branch. The new Korean Constitution, which took effect the day Roh Tae Woo assumed the nation's highest office, provides the National Assembly with increased powers and a more significant role. The National Assembly has the power to enquire into government affairs and may use it to investigate ex-president Chun and the Kwangju incident. However, the President retains many of the sweeping powers held by his predecessors (though not the power to dissolve the National Assembly). Thus, Roh and his cabinet will continue to play a dominant role in making and administering government policy.

Ex-president Chun Doo Hwan

Ex-president Chun Doo Hwan is nearly 57 and a former lieutenant general. He was born into a farming family in Hapchon, South Kyongsang Province and, like Roh Tae Woo, was a regimental commander in Vietnam (1970-72). The ex-president's role in the government and Korean politics remains an unanswered question. After leaving office, Chun joined a quasi-public organization called *The Advisory Council of Elder Statesmen*. Shortly before leaving office, he gave this organization quasi-public status and the power to require that government ministers consider its advice. Chun also retained his position as honorary chairman of the ruling DJP.

Two weeks prior to the National Assembly elections, Chun announced that he was resigning from his position with the Advisory Council of Elder Statesmen and his honorary chairmanship of the DJP, although he will remain a regular member of the DJP. Also resigning from the Advisory Council were several long-time Chun loyalists. Many believe Chun's announcement was part of a deal under which the government agreed not to prosecute Chun himself in return for the resignations. The Roh administration has publicly rejected these deal allegations. The allegations which began to surface in early 1988 concerning Chun's younger brother (Chun Kyung Hwan) are another indication that the influence of the ex-president has been substantially reduced.

The developments involving charges brought against his younger brother are both a threat and an opportunity for the current Roh administration. The threat is that the scandal will spread to other officials in the Roh government, the DJP or the military establishment. To forestall any potential criticism, Roh has publicly announced that he will not appoint any relatives to public office. The ex-president has called for a full investigation of the charges against his brother.

Some political observers believe that the allegations against Chun's brother were leaked intentionally by DJP officials in an attempt to consolidate the power of the Roh administration. Supporting this theory was the DJP's decision to oust many of Chun's supporters from the party's slate of candidates for the April 1988 National Assembly elections.

Allegations have also begun to surface concerning ex-President Chun himself. It must be remembered that during his adminstration Chun constructed a relatively imperial presidency which matched his ego and he may have expected much of his power and perquisites to continue after relinquishing the presidency. Sources have indicated that President Roh has refused to accept Chun's proposals concerning his continued participation in government and that the ex-president has become very irritated as a result.

Other political power groups

Whatever the nature of ex-President Chun's role will be, the military will continue to be a politically powerful force in Korea. The Korean military exercises its political influence in many ways. Most importantly, ex-officers hold many key government and business positions. Korean generals have a history of threatening "stern actions" against individuals whose policies would harm the military or who are believed to be pro-communist. The involvement of the military in politics has a long history in Korea and this involvement cannot be expected to change in the foreseeable future.

Another major factor in the political scene in Korea are a group generally referred to as the *students*. In reality this group includes unemployed college and high school graduates, the underemployed (people working in jobs for which they are over-qualified), the economically disenfranchised and the political left as well as college students. This group has disproportionate visibility because of the tradition of student unrest in Korea. It was student-led unrest that brought down the government of President Syngman Rhee in 1960. Students also played an important role in resisting the Japanese colonial occupation of Korea. Largely because of the influence of Confucian teachings, students are expected to be the conscience of the nation. In fact, students have a desire for political change but little knowledge about what specific changes they would implement. In general, the students are extremely nationalistic and suspicious of foreign powers. It is significant to note that the students have not personally suffered Japanese colonial occupation or experienced the benefits of American military involvement. They are significantly more radical than their parents, less afraid of Japan and less impressed by western ideas. Two major goals of the students are reunification of the nation and a civilian government.

Recently, Christian church groups have become a major factor in domestic politics. Over 20% of Korea's population is Christian and an even larger number look to Christian church leaders for guidance. Church groups in Korea are well organized (particularly in urban areas) and have effective leaders. In the past, the Catholic Church (which has approximately two million members) has been more politically radical than Protestant denominations (which have collectively many more members than the Catholic Church). The Korean Catholic Church is headed by Cardinal Steven Kim Sou Hwan. In general, social welfare goals and democratization are the primary focuses of the church groups' political activities. In many cases, church leaders have been effective moderators in government-student and government-labour confrontations.

Korean professionals and intellectuals form another pressure group. These relatively western-oriented individuals can widely influence public opinion despite their small numbers. The major task facing professionals and intellectuals is to assimilate western ideas and technology without losing their character as Koreans.

Neither the students nor the Christian Church can effect political change in Korea without the support of the middle class. This group was largely responsible for the political changes which occurred in the summer and fall of 1987. The middle class is slow to move as a group but when moving can effect major change. The key to keeping this group's support is continued economic progress. As long as the middle class feels their economic well-being is improving, they will continue to accept a lack of political freedom. However, as President Park found during the economic downturn in 1979, the middle class is fickle in its political support.

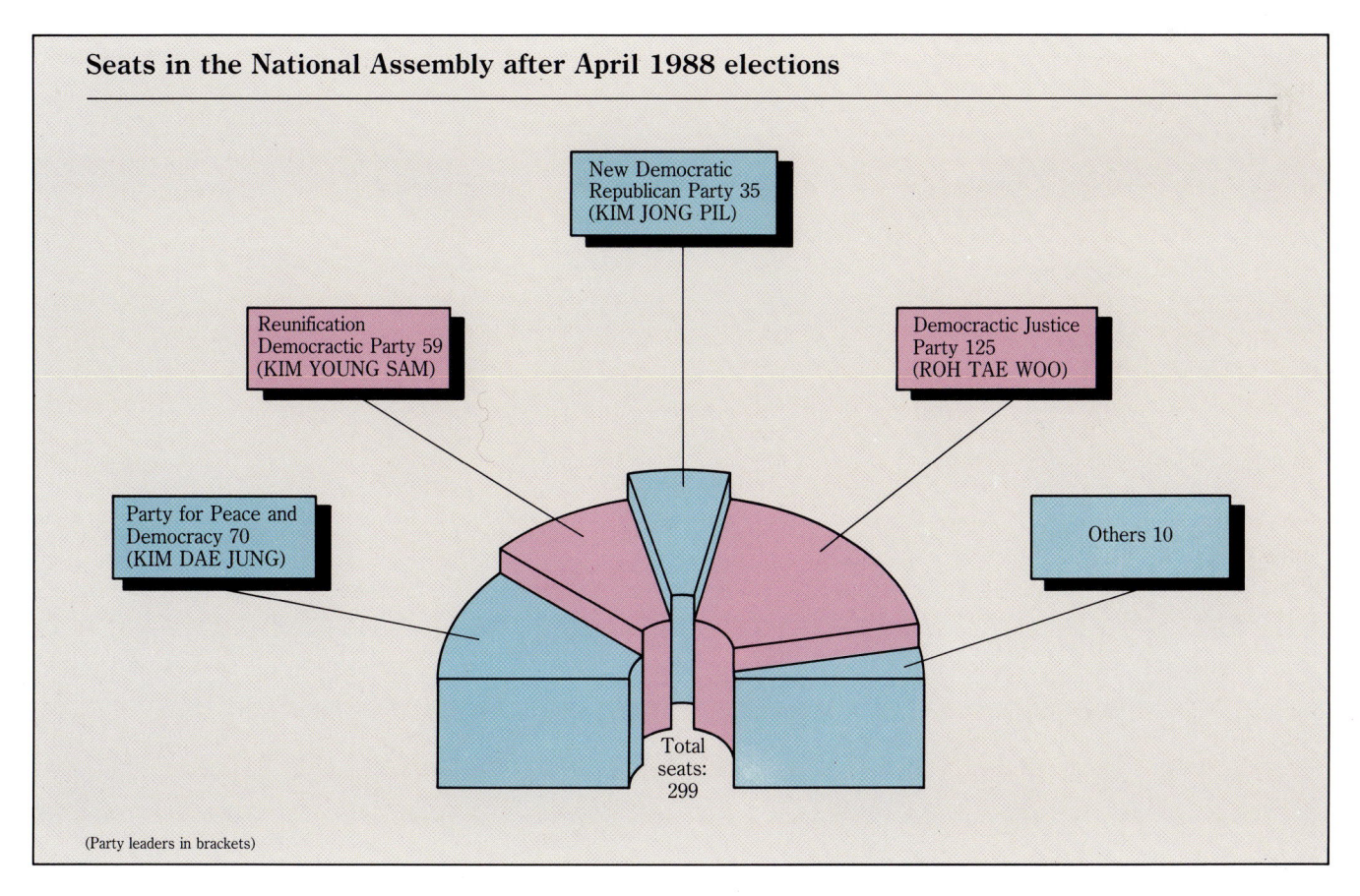

Seats in the National Assembly after April 1988 elections

New Democractic Republican Party 35 (KIM JONG PIL)

Reunification Democractic Party 59 (KIM YOUNG SAM)

Democractic Justice Party 125 (ROH TAE WOO)

Party for Peace and Democracy 70 (KIM DAE JUNG)

Others 10

Total seats: 299

(Party leaders in brackets)

The domestic and overseas construction industry has played an important part in the development of the Korean economy

Chamsil Olympic stadium – home of the 1988 Olympic games

Chapter VI
The economy

Exports and the three blessings

Korea's economic miracle has been a remarkable achievement. When Korea adopted the first Five-Year Economic Development Plan in 1962, per capita income was US$82 and total exports only US$43 million. Domestic savings and investment were low and the economy was principally agrarian. In 1987, per capita income had risen to US$2,850 and exports were approximately US$47.28 billion. The principal engine of this economic growth has been exports. The Korean government has established an impressive set of policy measures to fuel this engine, including subsidies, special financings and tax incentives.

Korea's recent economic growth is said to be due to the *three blessings* – low oil prices, low interest rates and the low value of the dollar relative to the yen. Korea imports all of its oil and gas, has a large external debt (approximately US$35 billion in 1987) and sells many products which directly compete with those made in Japan (e.g., steel, ships, video cassette recorders, televisions and automobiles). Even without the three blessings, Korea's low inflation, high domestic savings rate, industrious workers and well planned industrial strategy would have produced impressive growth. In years to come, Korea's ability to continue its spectacular economic gains will be limited by the *three highs*-higher value of the won, higher labour costs and higher raw material costs.

Korea's industrial sectors

Manufacturing

Since the first EDP was adopted, Korean manufacturing industry has developed in three different stages. From 1961–72, the development of light industries with a large labour content was emphasized. These light industries included textiles, apparel and footwear. By the end of the 1960s, Korean light industry was a large exporter and a significant factor in this sector of the world economy. As Korea's economy began to develop and lose its comparative labour cost advantage, and as the need for intermediary raw materials (such as rubber and plastics) increased, Korea's second stage, the development of heavy and chemical industries, began to surge. The government supported this change of emphasis in the third and fourth EDPs. Between 1965–80 the heavy and chemical industries increased as a proportion of manufacturing activity from 34.2% to 53.2%. Korean economic planners emphasized heavy industries such as steel production and shipbuilding during this period.

Korea's move toward heavy industry was accelerated by bilateral restraints imposed by the US on textiles and apparel. Much of the justification for the heavy industry development policy was based on an import substitution theory. Heavy industry was also thought to have strong links to light industry. The *strategic heavy and chemical industries* received preferential financing and protection from imports. Korean companies in this sector were allowed monopolistic production and pricing policies. The move of investment funds to this sector was so large and fast that 71% of investments in the manufacturing sector went to this area from 1973–76, and 79% from 1977–79. Excess capacity combined with surging oil prices created a crisis in this industry in 1980.

In the early 1980s, the manufacturing industry in Korea began to move into a third stage in which technology-intensive manufacturing began to play the leading role. These higher technology industries should overtake both light industry and heavy and chemical industry by 1990 as the leading manufacturing category. Included in this rapidly developing category are the electronics, automobile, precision machinery and computer industries.

Part of the motivation for this change in industrial structure was the imbalance caused by the excessive investment in heavy and chemical industries. The number and influence of small and medium companies had been substantially reduced by the economic strategy favouring the heavy and chemical sector. This lack of smaller businesses meant that Korea's manufacturing industry was handicapped in its ability to produce parts and components and to produce machinery and electronics. The nation's balance of payments (and in particular its trade deficit with Japan) suffered as a result. Furthermore, the country's manufacturing technology had become seriously out of date. In the early 1980s steps were taken to encourage a transition toward more capital and technology intensive industries. By 1986, Korean industry was moving quickly toward the standards of developed industrial countries in the high technology sector.

Shipbuilding. This industry played an important part in Korea's rapid economic development and was targeted as a key industry in the third EDP because of Korea's comparative advantages and because of its links to other key industries (steel, machinery and electronics). From 1977–83,

Korean shipbuilding capacity increased from a very low level to take a 20% share of the world market. In 1986 Korea surpassed Japan as the world's largest holder of orders for vessels. Japan and Korea held equal orders in 1987 with Korea's share totalling some 3.4 million g/t in 1987, an increase of approximately 28% over 1986. The Korea Shipbuilding Association predicts orders will reach 3.7 million g/t in 1988. This would give Korea 25% of the world shipbuilding market. The backlog at the end of 1987 was 5.8 million g/t.

Recent double figure wage rises in this industry have made Korea less competitive in relation to the People's Republic of China, Brazil, Thailand, India and other developing countries. The major weakness of the Korean shipbuilding industry is its continued reliance on imported intermediary goods. Only 56% of parts and components used in building ships in Korean shipyards are produced in Korea. Recent increases in the value of the Japanese yen have increased the price competitiveness of the Korean shipbuilding industry in relation to its biggest competitor. However, since export financing provided by the Korean Export Import Bank is based on the won, substantial foreign exchange losses will accrue to Korean shipbuilders as the value of the currency continues to appreciate.

The major Korean shipbuilders are Hyundai Heavy Industries, Daewoo Shipbuilding, Samsung Shipbuilding and Korea Shipbuilding and Engineering. The industry suffered a major blow when Korea Shipbuilding and Engineering was placed under court receivership raising concerns that foreign banks may not continue their financial support of this industry if they incur substantial losses in this receivership. The Korean shipbuilding industry has very high debt levels: total debt at the end of 1986 was US$105.1 million.

The Korean shipbuilding industry has been affected negatively in recent years by falling demand for ships and world over-capacity for shipbuilding. Labour disputes in 1987 and resulting wage increases and delivery delays also hurt the industry. Nevertheless, Korea should be able to maintain its current share of the market and to increase localization of ship components. Korean shipbuilders will continue their efforts to produce more special purpose, technically sophisticated vessels with higher value added content.

The Korean government will also work to increase domestic demand for ships. The government's *Planned Shipbuilding Programme* will continue to offer below-market interest rates and favourable terms to domestic shipping companies that purchase Korean vessels. As is the case in other industries, the government is working to eliminate excessive competition between Korean companies.

The profitability of Korean shipbuilding companies should increase in coming years as the global shipping industry begins to recover. However, former production levels and industry profits may not be reached until the mid-1990s. In the long-term, shipbuilding is a sunset industry in Korea.

Automobiles. The automobile industry stretches back to the mid-1940s with the assembly of auto parts and knockdown components. For 20 years this hardly changed, except in scale. The industry has experienced incredible growth since the first assembly plant was established in 1962. This industry has gone through three distinct stages: (1) assembly of imported parts *1962–65*; (2) assembly of imported parts and limited component manufacturing *1965–74*; (3) manufacturing of Korean-made vehicles with importation of key components *1974-present*. Total automobile production in 1987 rose to 795,802 units. This rapidly growing sector of Korea's economy should continue to experience significant growth both on domestic and export markets.

The three major exporters of automobiles are Hyundai Motor, Daewoo Motor Company (a joint venture with General Motors) and Kia Motors Corporation (in which both Ford Motor Co. and Mazda have a minority stake). It is estimated that Hyundai sold 400,000 cars in 1987 in export markets. In 1987, Daewoo and Kia also exported 90,000 and 65,000 cars respectively. These exports had a value of around US$2.7 billion. Two other Korean transportation equipment companies (Dong-A and Asia) produce vehicles for domestic use. To date, most exports are in the small car category. Hyundai has been selling the *Excel* in the Australian, US, and Canadian markets with great success and should soon move to sell a version of the more upscale *Stellar* on the US market (the Stellar has already been sold in Canada). Hyundai will soon introduce its mid-size Y2 class of cars in order to broaden its product range. Daewoo has been exporting the *Le Mans* to General Motors for sale in the US under the Pontiac name. Kia produces a minicar with technology supplied by Ford. Kia's exports are sold domestically as the *Pride* and overseas as the *Festiva*. In addition to purchasing finished automobiles, foreign automobile manufacturers will continue to purchase a larger share of their parts and components from Korea.

The three car-exporting companies have set a combined export target of approximately 700,000 vehicles for 1988. Exports of automobiles in the first quarter of 1988 were US$856.4 million, an increase of 102.4% over the same period in 1987. Domestic sales are projected to rise to 500,000 units in 1988. Domestic demand for cars should continue to rise as the standard of living in Korea increases. Global production over-capacity for automobiles means that Korean firms can continue to be successful in export markets only if prices are kept down and quality up.

Labour disputes could reduce profitability in this industry during 1988. A strike at Daewoo Motors substantially reduced production in 1988. It is estimated that labour unrest during 1987 prevented the production of around 55,000 cars. To buffer themselves from increasing labour costs, facility investments will increase dramatically. All Korean manufacturers are increasing research and development expenditures to meet increasing competition. To counter protectionist moves, Hyundai will begin producing cars in Canada in early 1989. Imports of approximately 1,000 to 2,000 foreign cars into Korea during 1988 and slightly more the next year should have little effect on industry profitability.

Consumer electronics. The consumer electronics industry began in Korea in the early 1960s when radio assembly lines were constructed to meet domestic demand. This industry has grown to become a huge contributor to employment and exports. Growth in this industry began to increase rapidly in the 1970s as production rose from US$138 million in 1971 to US$5,558 million in 1983. The factors contributing to this success include active government support, tax benefits for foreign investors, a skilled but cheap workforce, and a growing domestic market. VCRs and colour TVs are expected to be the export market leaders with a growth of around 30% annually for the next five years. The Korean government has fostered capital and technology contributions to the electronics industry from abroad by encouraging joint ventures. Recent annual growth in this industry has exceeded 30%. Exports of electronics from Korea had a value of approximately W$10 billion in 1987, making Korea the sixth largest exporter of this type of product. In 1988, exports of electronics are expected to exceed US$13 billion. Korea currently has approximately a 5% share of the international electronics export market. During 1988 or 1989 electronic exports should exceed those of textiles to make it Korea's leading export industry. By 1992, exports of electronic products may reach US$25 billion. The Korea Institute for Economics and Technology has issued a study predicting that electronics industry will represent 17.2% of all manufacturing and 25% of exports in the year 2000.

A broad range of electronic products are manufactured by Korea. The largest Korean companies in this industry are Gold Star, Daewoo Electronics and Samsung Electronics. The combined production by these three companies is expected to double to reach approximately W10 trillion by 1990. Many small and medium sized companies are engaged in producing electronic goods as well as parts and components.

The principal weakness of the electronics industry is its reliance on Japanese suppliers for key parts and components. In terms of value, 40% of all components used by the industry are sourced from Japan. By 1990 this figure should be reduced to 25%. Increased localization will result in an increase in profitability for Korean electronics companies. Another problem with the electronics industry is its excessive reliance on the US market. The contribution of wages to total production costs is lower in the electronics industry than in some other industries, so recent wage increases should not be as harmful as in some other cases. However, the domestic market for electronic products should continue to increase in size as per capita income increases. Export sales should also continue to increase, although not as fast as in recent years. Profitability in this industry should continue to be very good.

Computers. Computer manufacture only began to move from simple assembly to product development in 1985. Exports were US$19 million in 1983, rising to US$395 in 1986. Total production in 1987 doubled to US$1.5 billion, but Korean manufacturers are still heavily dependent on Japanese components and foreign software to substantially increase their market share. Korean computer companies have captured a large share of the world market for personal computers and computer parts and components. In 1987, the computer industry exported US$380 million and accounted for 10% of Korea's electronic exports. Korean companies have to date been most successful in producing personal computers for sale bearing foreign trademarks. Daewoo's production of the *Leading Edge* personal computer is the most notable example. More and more products will bear Korean-owned trademarks in years to come. Hyundai and Samsung have recently begun selling an IBM-clone personal computer bearing their own name in the US and other markets.

It is doubtful whether Korea can begin to produce mainframe computers in the foreseeable future. The development of Korean software for both personal computers and mainframes is still at an infant stage. Most efforts will be directed toward adopting foreign software for the domestic market. Korean manufacturers have made huge investments in semiconductor manufacturing facilities and should capture an increased share of this market. As a rule, Korea's principal contribution to the semiconductor market will take the form of generic rather than custom microchips.

In the years to come, increasing world demand should cause a steady rise in the profitability of Korean computer companies.

Metals. Despite the fact that Korea has no iron ore and is located far from raw material sources, Korea's economic planners decided at an early stage to establish a large-scale iron and steel industry. The most important component of the plan was the decision to establish an integrated iron and steel complex at Kwangyang Bay in Pohang, a port city in southeast Korea. These facilities are operated by the Pohang Iron and Steel Company (POSCO), a mainly government owned entity. The production facilities for iron and steel established by POSCO as well as other Korean companies have made Korea one of the world's ten largest producers of crude steel. With the completion of the second construction phase, total capacity at POSCO's facilities is 14.5 million metric tons. If that much steel is actually produced, POSCO will be the second largest steelmaker in the non-communist world (after Nippon Steel but ahead of USX). POSCO has announced a plan to build a third production facility at Kwangyang which will add 2.7 million tons of capacity. This third phase should be completed in 1991. To complete this latest facility, an investment of US$2.32 billion will be needed.

POSCO and USX have entered into a successful joint venture to export steel to the US. The profit from this venture increased to US$15.87 million in 1987 on a turnover of US$547.58 million. One-third of Korea's steel output is typically exported to the US and Japan. Exports of steel products in 1987 increased by approximately 7% over the previous year. In the future, Japan should replace the US as Korea's largest export customer. Exports of steel should decrease relative to domestic consumption since the demand for steel by local companies is increasing rapidly. Experts predict that demand for steel should exceed supply in Korea until the mid-1990s. Domestic sales in 1987 increased by 20% over 1986 and should further expand by approximately 12% in 1988.

While production of non-ferrous metals is increasing in Korea, this product will never be a significant export industry. Imports of non-ferrous metals will always be necessary to satisfy domestic demands.

The long- and short-term profit outlook for the Korean metals industry is good since demand for steel by domestic automobile and electronics companies is increasing rapidly. This increase in demand should more than offset raising wage and raw materials prices.

Machinery. Thanks to government support for heavy industry, the industrial machinery sector recorded high growth rates from the mid-1960s, reaching a production level of US$1.47 billion in 1986 and exporting US$890.8 million. Office machinery was another high demand area, with a significant US$261 million being imported to satisfy demand (up by 78.8% over 1985), and US$228.2 million in export sales. Production of industrial machinery and equipment in Korea has enjoyed rapid growth in recent years, particularly in domestic markets. The most significant weakness of this industry is that the quality of Korean machinery continues to be lower than that of imported products. However, for machinery utilizing low and medium technology, Korean products are very price competitive in domestic and export markets. Japan and Germany (and to a lesser extent the US) have captured a good share of the Korean domestic market for machinery by concentrating on the premium sector. Continued Korean government restrictions on imports, production subsidies, tax benefits for users and increasing domestic demand (nearly 2.5% per year) should result in continued profits for this industry.

The most exciting aspect of the machine tool industry is the development of numerically controlled (NC) machines because they are more precise, productive, and easily programmed to reflect changes in production requirements. In 1986, NC machine tools accounted for 50.2% of export value. Production as a whole is expected to be US$390 million in 1988, and US$480 million in 1989.

The precision instruments industry, comprising timepieces, optical instruments, measuring apparatus and medical appliances, had expanded by an average of 40% per year between 1970–86. The growth was impressive but technology and quality are not very high. Nearly two-thirds of domestic demand was imported in 1986.

Petrochemicals. The Korean government's economic planners have targeted the petrochemical industry for development in each of Korea's economic plans (particularly in the third and fourth plans). Korea reached self-sufficiency in the production of chemical fertilizers in the late 1960s and is working toward self-sufficiency in other chemical sectors. Numerous refining and processing facilities, including naphtha crackers and aromatic extraction plants have been constructed in Korea as part of this effort. Some of the major companies involved in the industry are Yukong, Honam, Korea Plastic, Kumho, Samkyung and Sunkyong.

Petrochemical refining and processing facilities in Korea are now operating at full capacity in nearly all sectors. The Korean Petrochemical Industry Association has reported that although production capacity in this industry for upstream products will increase to 3.5 million tons during 1988, shortages of some products (particularly synthetic resins) are expected. Fortunately, there are approximately 30 expansion projects which will nearly double existing capacity due to come on line over the next few years.

Profitability in this industry is heavily dependent upon world oil prices, since 100% of the raw materials are imported. The existence of *forward linkages* between the petrochemical sector to rapidly growing industries like electronics and automobiles should result in continued profitability.

Growth in the petrochemical industry should exceed 11% in 1988. Total production of petrochemical products rose 13.9% to 2.214 million tons in 1987. Since domestic demand is increasing at a pace approaching 15%, exports are falling as a proportion of total sales.

Toys. Until recently, the toy industry in Korea was relatively decentralized, with many small contractors collectively forming a significant presence in the international toy industry. Exports of stuffed toys were the principal focus of these companies, representing 73% of total toy exports in 1987. In recent years, Korean toy makers have been producing more electronic toys containing integrated circuits. As a result, the size of companies in this industry has grown to allow for the more sophisticated assembly lines to be set up. Even so, over two-thirds of all Korean toy makers are small companies. Exports of toys from Korea have grown by an average of 50% over the past three years, exceeding US$1 billion in 1987.

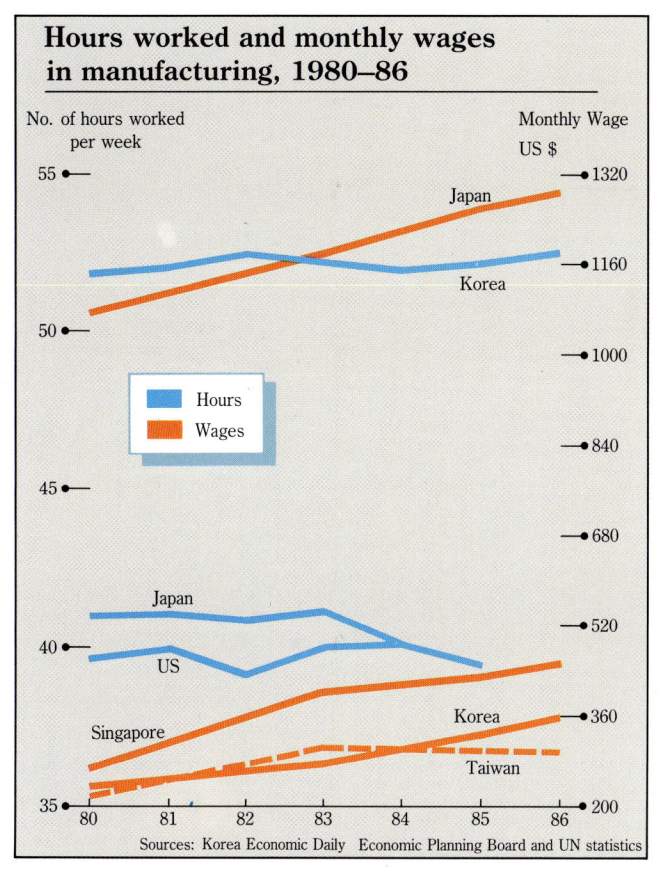

Hours worked and monthly wages in manufacturing, 1980–86

No. of hours worked per week / Monthly Wage US $

- Japan — 1320
- Korea — 1160
- 1000
- 840
- 680
- 520
- Korea — 360
- Taiwan
- 200

Lines: Japan, US, Singapore (Hours); Japan, Korea, Singapore, Taiwan (Wages)

Sources: Korea Economic Daily, Economic Planning Board and UN statistics

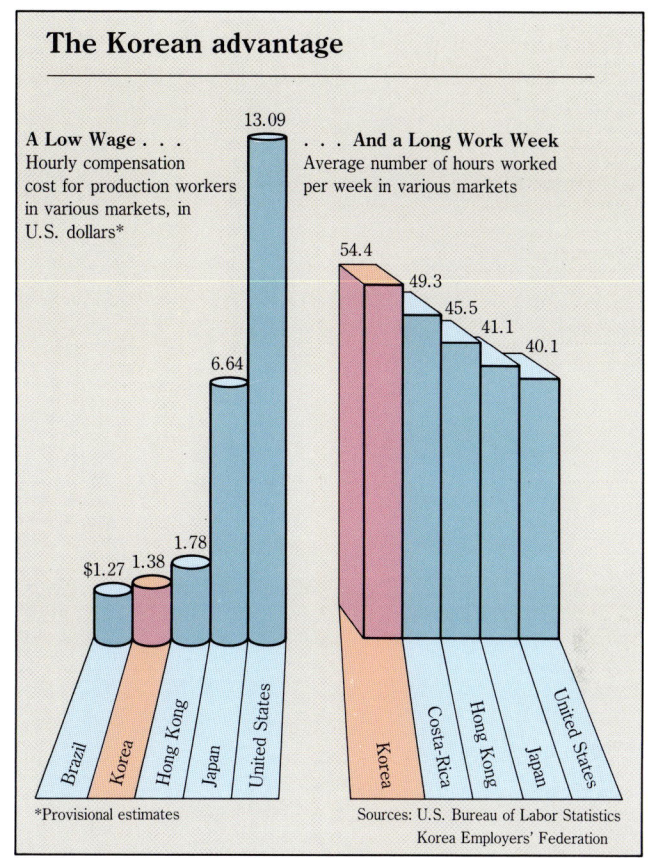

The Korean advantage

A Low Wage...
Hourly compensation cost for production workers in various markets, in U.S. dollars*

- Brazil: $1.27
- Korea: 1.38
- Hong Kong: 1.78
- Japan: 6.64
- United States: 13.09

...And a Long Work Week
Average number of hours worked per week in various markets

- Korea: 54.4
- Costa-Rica: 49.3
- Hong Kong: 45.5
- Japan: 41.1
- United States: 40.1

*Provisional estimates

Sources: U.S. Bureau of Labor Statistics, Korea Employers' Federation

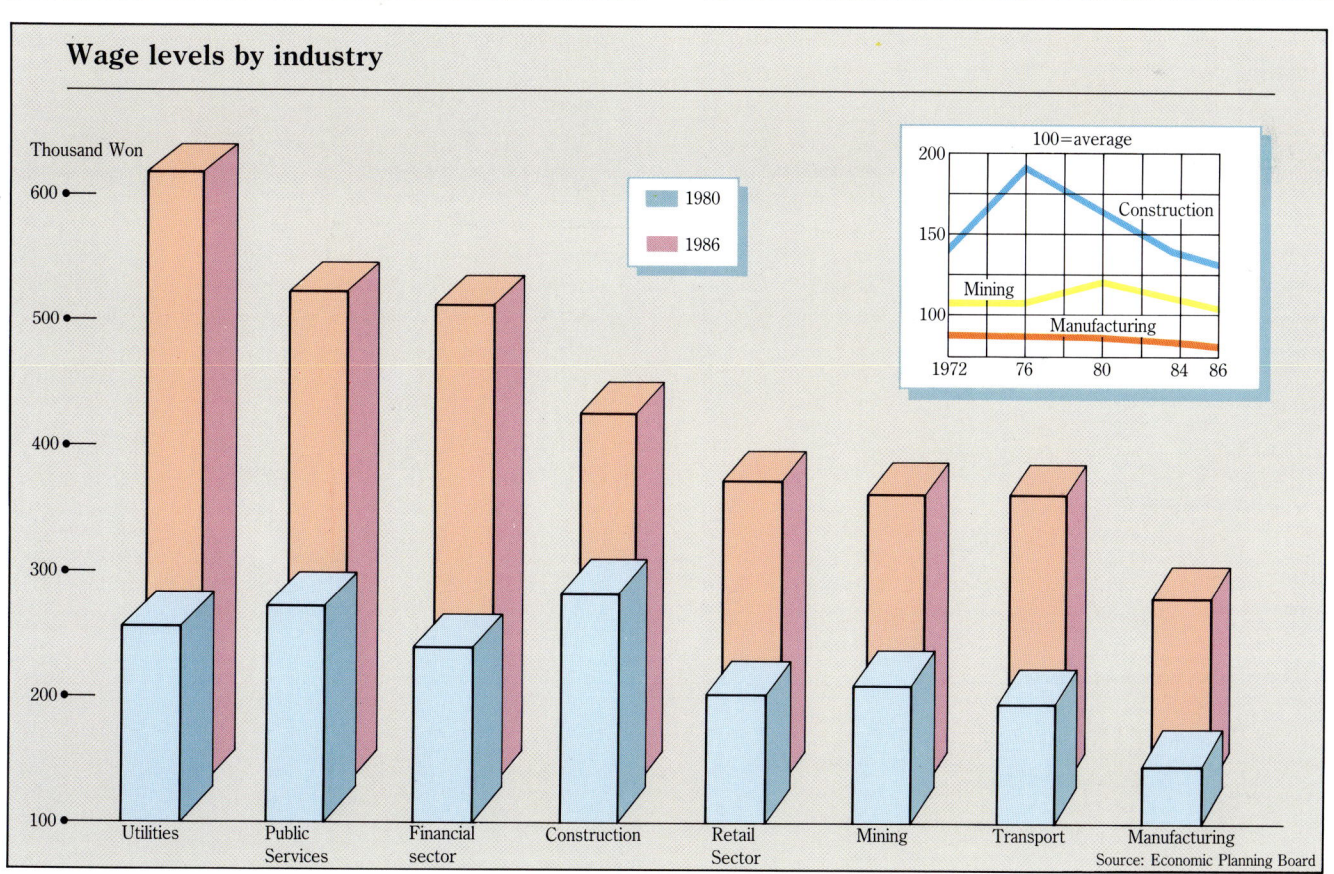

Wage levels by industry

Thousand Won

Legend: 1980, 1986

Industries (1980 vs 1986): Utilities, Public Services, Financial sector, Construction, Retail Sector, Mining, Transport, Manufacturing

Inset (100 = average):
- Construction
- Mining
- Manufacturing
Years: 1972, 76, 80, 84, 86

Source: Economic Planning Board

Export and import trends

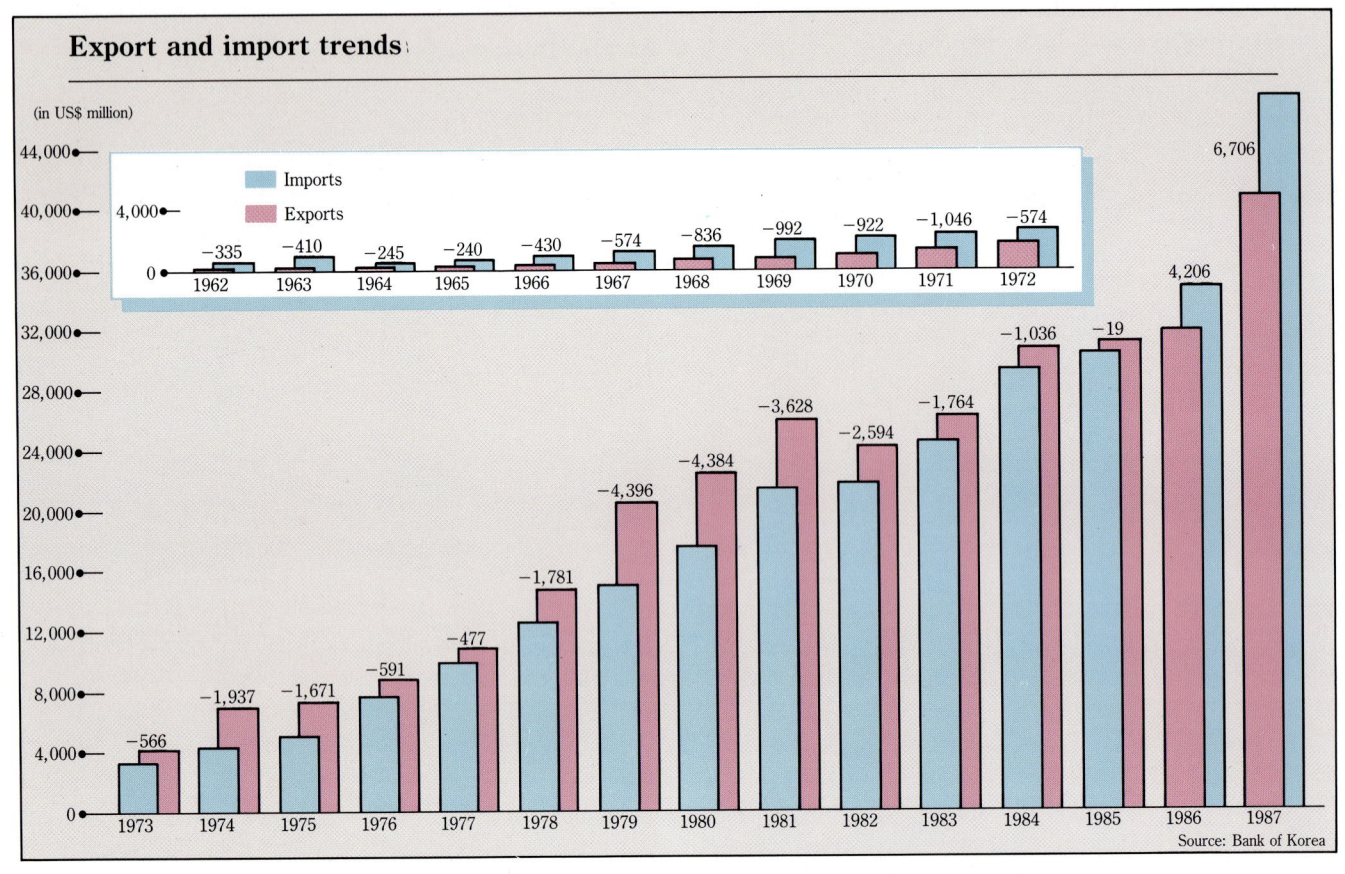

(in US$ million)

■ Imports
■ Exports

Year	Balance
1962	−335
1963	−410
1964	−245
1965	−240
1966	−430
1967	−574
1968	−836
1969	−992
1970	−922
1971	−1,046
1972	−574
1973	−566
1974	−1,937
1975	−1,671
1976	−591
1977	−477
1978	−1,781
1979	−4,396
1980	−4,384
1981	−3,628
1982	−2,594
1983	−1,764
1984	−1,036
1985	−19
1986	4,206
1987	6,706

Source: Bank of Korea

Korea's export markets in 1987

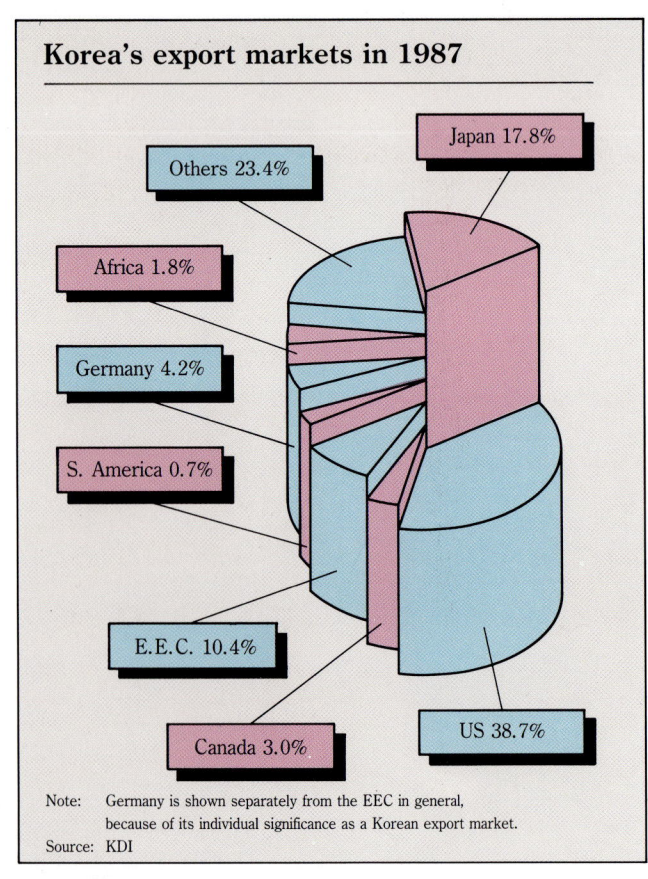

- Japan 17.8%
- Others 23.4%
- Africa 1.8%
- Germany 4.2%
- S. America 0.7%
- E.E.C. 10.4%
- Canada 3.0%
- US 38.7%

Note: Germany is shown separately from the EEC in general, because of its individual significance as a Korean export market.
Source: KDI

Korea's main exports

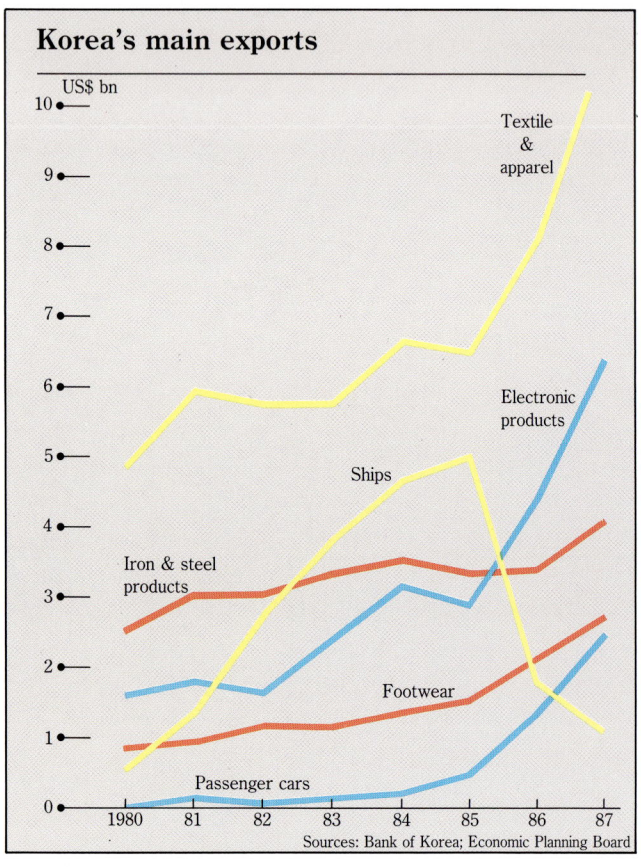

US$ bn

- Textile & apparel
- Electronic products
- Ships
- Iron & steel products
- Footwear
- Passenger cars

1980–87

Sources: Bank of Korea; Economic Planning Board

Current account balances, 1977–1987

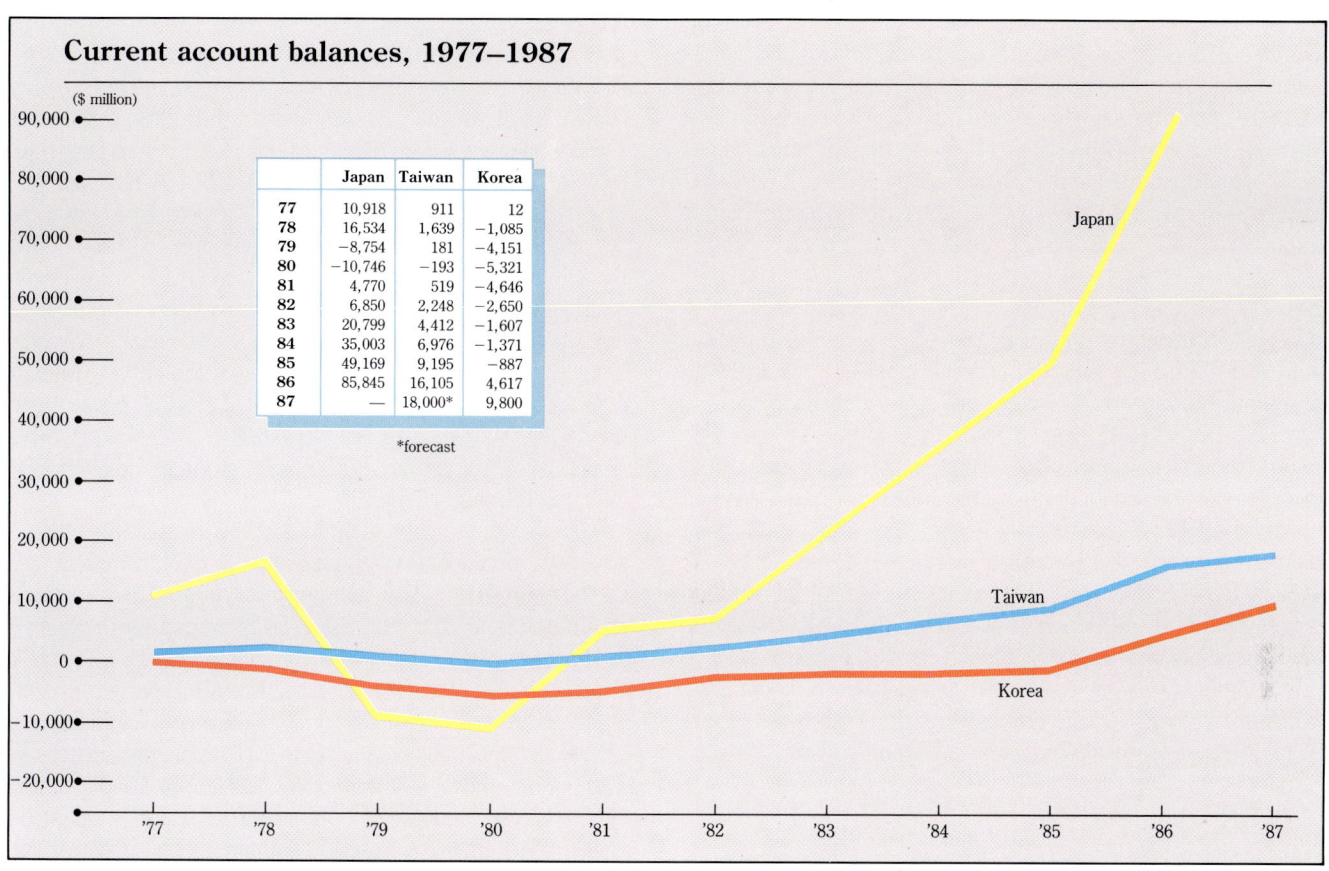

($ million)

	Japan	Taiwan	Korea
77	10,918	911	12
78	16,534	1,639	−1,085
79	−8,754	181	−4,151
80	−10,746	−193	−5,321
81	4,770	519	−4,646
82	6,850	2,248	−2,650
83	20,799	4,412	−1,607
84	35,003	6,976	−1,371
85	49,169	9,195	−887
86	85,845	16,105	4,617
87	—	18,000*	9,800

*forecast

The Five Year Economic Development Plans (EDP)

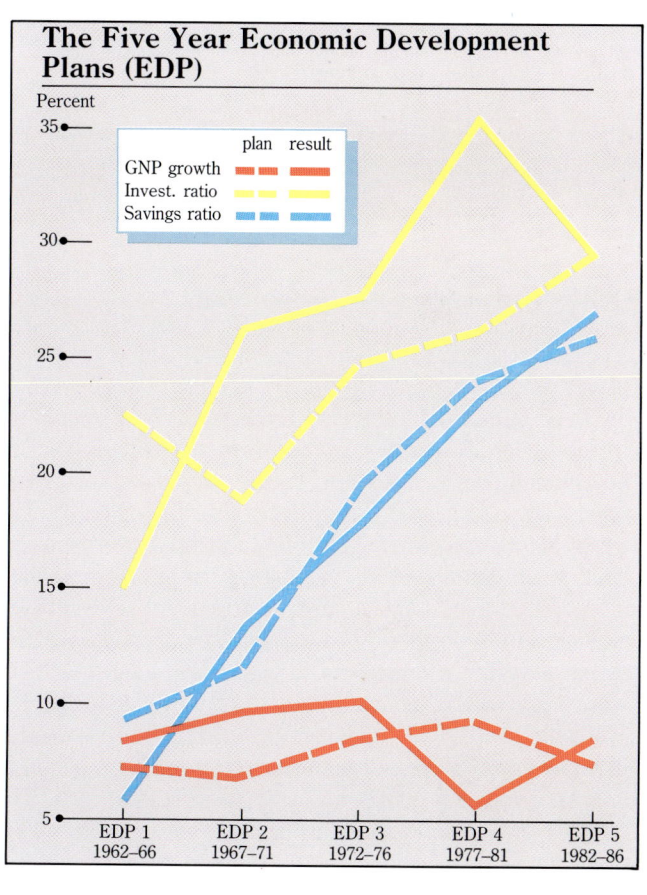

Percent

- GNP growth (plan / result)
- Invest. ratio (plan / result)
- Savings ratio (plan / result)

EDP 1 1962–66
EDP 2 1967–71
EDP 3 1972–76
EDP 4 1977–81
EDP 5 1982–86

The Korean and Japanese population structure (1986)

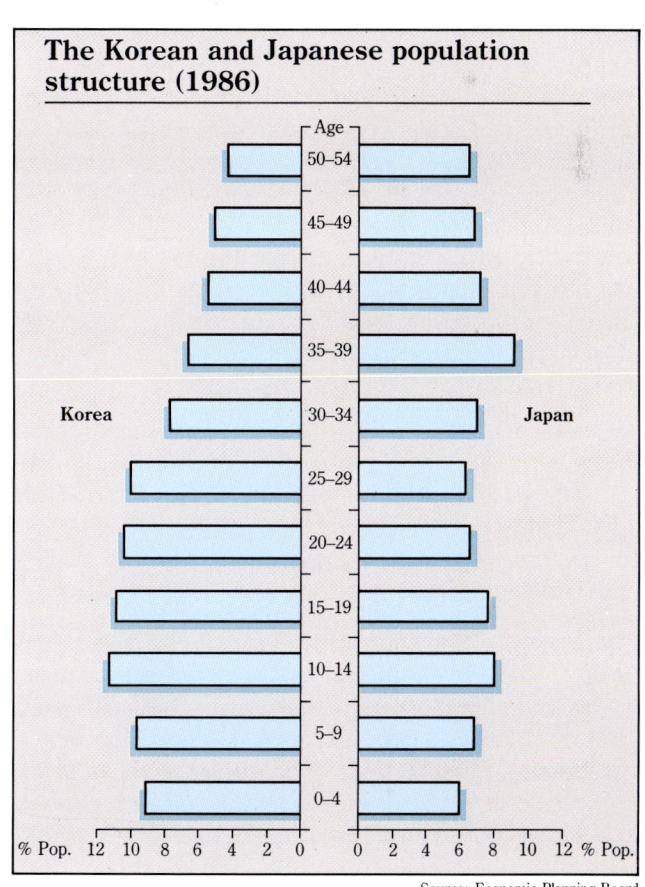

Korea / Japan

Age groups: 0–4, 5–9, 10–14, 15–19, 20–24, 25–29, 30–34, 35–39, 40–44, 45–49, 50–54

% Pop. 12 10 8 6 4 2 0   0 2 4 6 8 10 12 % Pop.

Source: Economic Planning Board

Textiles. The textile industry has been a major factor in Korea's economic development since the first export shipments of cotton garments in 1957. Until recently, the textile industry was the single most important contributor to exports, employment and the balance of payments. The development of this industry has enabled Korea to become one of the world's five largest textile producers. Total exports in 1987 were US$11.7 billion, an increase of 34% over the previous year, and accounted for 70% of sales. Exports are expected to reach US$13 billion in 1988. To head off protectionist moves by the US, the textile industry will work hard to reduce dependence upon the US market (currently taking around one-third of Korean exports) by increasing exports to Japan. Korea has traditionally devoted most production capacity to synthetic fibre garments, but increases in the production of natural fibre garments will be necessary. While the prominence of this industry in the Korean economy will decrease in relative terms, textiles will continue to be very important. However, increasing wages paid to Korean textile workers, the rising value of the Korean won and protectionism in world markets may prevent any significant profit and sales gains in this industry. Some growth will occur in domestic markets but not enough to fully compensate the industry for export losses.

Footwear. The footwear industry was originally established in 1920 to satisfy local demand and expanded by virtue of its skilled and low-cost labour force The footwear industry has been a significant part of the Korean economy for over 20 years, and Korea is now one of the world's largest producers of footwear. The four largest shoe manufacturers in Korea are HS Corporation, Kukje, Samwha Company and Tae Hwa Corporation. However, the footwear market is also supplied by a large number of small and medium companies. Korea shipped footwear with a value of US$2.1 billion to export markets in 1986, making this industry Korea's fourth largest export industry. Exports in 1987 were approximately US$2.8 billion, an increase of over 30% over 1986. Most of the footwear exported from Korea are leather-upper athletic shoes.

Despite recent growth, a major weakness exists since the footwear industry is very dependent upon the US market. The People's Republic of China, India, Thailand and other third world countries are becoming increasingly tough competitors in this market. Increased automation is planned by the Korean footwear industry to meet this competition.

Construction. The construction industry is a highly competitive, aggressively international industry facing a broadly favourable future. Successive EDPs have provided technical and financial aid and direction for the industry. The ratio of construction projects to GNP is stable at about 8–9%. Domestic construction is helping to make up for the decline in overseas orders caused by falling oil prices. The domestic and overseas construction industry has played an important part in the development of the Korean economy. Most of the success of Korean construction companies derives from the Middle East market with gross earnings rising to nearly US$15 billion in 1978. These Middle East construction projects provided employment for over 122,000 Koreans at its peak. Without the foreign currency earnings from this industry, the Korean economy would have suffered greatly during the oil-crisis years. Since the decline in OPEC's power over oil prices, the volume of overseas construction projects has decreased dramatically. To make matters worse, payments from the Middle East have been slow in coming. Ministry of Construction officials announced in 1988 that of the US$83 billion in Middle East construction work performed by Korean companies since 1968, payment of only US$70.7 billion had been received.

Many Korean construction companies have had to severely retrench or combine with healthier companies due to the slowdown. The Korean government has, in typical fashion, rationalized the overseas construction industry by licensing only 15 financially healthy companies to serve as contractors on overseas projects. Korean construction companies have only been marginally successful in finding new overseas markets. New orders in overseas markets fell 22.7% to US$1.7 billion in 1987. The only good news for 1987 was the Ministry of Construction's announcement that Korean overseas construction companies collectively registered a surplus of US$87 million, the first in the 1980s. For this reason, this industry will be largely driven by domestic demand from this point forward. The construction industry will be stimulated by increased public works projects in Korea, particularly in provincial areas. However, the 1986 Asian Games and the 1988 Olympics have boosted the construction industry as well as domestic demand for cement and plate glass.

Other industries and resources

Fisheries. Although inland and coastal fishing have long played an important part in Korea's economy, deep-sea fishing only dates from 1957 with the first catch of tuna in the Indian Ocean. Korea now has 800 deep-sea vessels catching pollack and pike in the North Pacific, octopus and squid off northwest Africa, and shrimp from Brazil. Coastal fishing, which amounts to nearly one-half of the industry's total production, is dominated by small-scale operations organized into cooperatives. These were instrumental in developing an aquaculture business (oysters and laver), producing 25% of the industry's tonnage. The production levels for 1986 were exceptional, some 3.7 million tons of fish and fish products worth US$1,384 million, but these are expected to show some decrease, as many countries have begun to impose swingeing quotas and sharply increased fees. The number of people working in Korea's coastal waters has

The Korean government has announced plans to double the income of farm householders to US$17,500 by 1993

The recently built Kwangyang steel works has the capacity to produce 14.5 million metric tons of steel

declined while employment in the deep-sea industry has increased markedly. To offset this loss of jobs, the government has taken steps to assist the growth of the domestic aquaculture industry. While the fishing industry has been profitable for the past 25 years, increases in wages and reduced fishing territories (as other nations enforce 200 mile offshore limits) create significant problems for Korea. The prognosis for the health of the fishing industry depends upon Korea's success in negotiating fishing rights agreements with other nations and on the success of the important *tsurimi* (processed pollock) industry. Joint ventures with companies from foreign countries will increase. The future for the fishing industry will lie in foreign joint ventures, bilateral pacts, replenishing fish stocks, and improved processing methods.

Mining. The Korean mining industry is severely hampered by a lack of domestic natural resources. Korea's mining activities are generally limited to the extraction of coal, tungsten, limestone, talc, silica, lead and zinc. Although anthracite coal is mined in significant quantities, Korea's production of coal cannot meet domestic demand. It is estimated that Korea has total coal reserves equal to 1.6 billion tons, but production costs are high and coal can be purchased from foreign sources for less. Nearly all of Korea's mining production is consumed domestically. To stabilize raw material supplies, overseas joint ventures in the mining area should increase significantly in coming years. Some resource extraction joint ventures have already been formed in the US, Canada, Australia and Indonesia.

Energy. Since Korea has no domestic sources of oil, it was necessary to make significant investments in power generating facilities. Increasing the supply of electric power was a major component of the early economic plans, largely the resposibility of the Korea Electric Power Corporation (KEPCO). Hydroelectric power generation was conducted through the Industrial Sites and Water Development Corporation. Major milestones include the commissioning of two nuclear plants in the 1970s and the completion of the rural electrification programme in 1978. During the 1982–85 period, four coal-fired plants, a pumped storage station and two LNG plants were commissioned. Power generating capacity in 1985 was a total of 16.137 megawatts, well above peak demand requirements.

The recent focus of Korea's energy policy is the replacment of petroleum power sources with nuclear and coal produced energy. Dependence upon imported petroleum dropped to 46.7% from 63.5% in 1978. As part of this programme, ten nuclear plants should be completed by the early 1990s. In 1986, 44% of total electricity was produced by nuclear energy. Coal-fired electricity generating plants are being constructed to replace oil burning plants. Efforts to increase the use of liquified petroleum gas (LPG) will also be made. LNG facilities will represent 8.6% of power generation capacity in 1991. Few additional hydroelectric sources in Korea remain to be tapped. The Ministry of Energy expects total energy demand in 1988 to increase by 5.1% from levels in 1987 to 70.5 million TOE. By 1991, total power generating capacity in Korea is projected to be 20,962 megawatts.

Agriculture. Approximately 60% of the land area of Korea consists of uncultivated and forested mountain slopes. Residential and industrial sites occupy another 10% leaving only 30% available for agriculture. That the mountainous terrain of Korea is unsuitable for pasture only exacerbates the problems facing agriculture. But agriculture remains a carefully protected sector for political and strategic reasons. There is a need to ensure basic food production, to forestall urban in-migration, and to balance industrial growth and rural decline/discontent especially since many leaders of the DJP are returned from rural constituencies. Farm prices in Korea are kept high (and therefore farmer's incomes) by forbidding the importation of competing products. Other products, such as grains, cotton, hides and skins form the bulk of the US$4 billion imported in 1986. As the cost of protectionism is borne by uncomplaining urban consumers rather than the government, the economic and political costs of liberalization seem unnecessarily high.

As there is so little land and since most farms are very small, Korea is unlikely to be self-sufficient in grain production. Korea may only be able to meet about 50% of its need for grain during 1988 even though over two-thirds of its farm land is devoted to producing grain. Half of this grain production area is planted with rice. This specialization has allowed Korea to become self sufficient in rice production. The two other important grain crops in Korea are barley and soybeans. Limited amounts of wheat and corn are grown.

Despite these inherent problems, Korea has been quite successful in developing its agricultural sector. Steady growth in food grain production has occurred during each of Korea's five-year economic plans. Gains have been particularly impressive considering Korea's traditional high productivity and the increasing conversion of farmland to industrial and residential sites. The principal focus of early efforts was on increasing the yield of grains by improving seed stock, fertilizer utilization and mechanization. While self-sufficiency of rice and barley was reached by 1977, imports of corn and wheat have risen dramatically.

Changes in food consumption patterns in metropolitan areas and increases in per capita income have caused increased demand for meat, fruits, vegetables and milk. Production of these items in Korea has in general risen to meet demand (and in the case of meat and milk, exceed demand). However, food prices are relatively high given Korea's level of development. The Korean livestock industry has high costs and must be protected from imports to survive. The US government will continue to pressure Korea to open its market for beef, cigarettes and alcoholic beverages. The only positive note for Korean farmers is their importance to the Democratic Justice Party. For this reason, prices paid by the government for rice and barley and certain other crops will continue to rise. Fruit producers may also expect continued protection from imports. The best hope for farmers is an increase in non-farm income, supported by the government.

The Korean government has announced plans to double the income of farm households to US$17,500 by 1993. The projections assume a 1.6% decrease in farm households, 3.5% yearly growth in the price of farm products and 3% growth in the agricultural sector.

Transportation. The Korean government has made an excellent job of expanding the transportation sector of the economy since the 1960s. As a whole, the increase in traffic volume has exceeded even Korea's impressive economic growth rate. This increase in transportation facilities was necessary, due to the poorly developed rail and road network left after the Korean War. Considerable future expenditure will be necessary to meet the growing demand for intra-Korea transportation. As a rule, export transportation is much more highly developed than its domestic counterpart. New port facilities are planned, particularly in the Cholla provinces. The prospects for profit in this sector are good in 1988 due to increases associated with the Olympics and increases in levels of trade.

Rail transportation. The railway network operated by the Korea National Railroad in 1985 covered 6,285 kilometres of track, 26% of which double-tracked and 14% electrified. The railway system carried 498 million passengers and 55 million tons of cargo in 1985. Despite huge growth in total cargo and passenger traffic, the rail system operates at a large deficit because of the substantial cost of investment and the government's policy of keeping fares low to benefit low income and rural individuals.

Vehicle transportation. The volume of land-based vehicular transportation in Korea has increased tremendously as the number of vehicles has skyrocketed. At the end of 1985, there were 1,113,430 motor vehicles registered in Korea. Of this total, 50% were cars, 12% were buses and 38% were trucks. Motorcycles are not included in these figures as they are not registered in the same way. As a consequence of this growth, major efforts will be directed at increasing the nation's network of highways and expressways. The nation's expressway system is managed by the Korean Highway Corporation which collects toll revenues and manages the roads.

Air transportation. The air transportation system in Korea has until recently been limited to one domestic carrier, Korean Air, part of the Hanjin business group, serving over 25 foreign airports and a large number of domestic cities. To increase competition, the government recently announced that the Kumho business group would be allowed to operate an airline in 1988 for the domestic sector, and then overseas in the early 1990s. This change ends Korean Air's 20 year monopoly on domestic routes. The Transportation Ministry announced in 1988 that the number of passengers on domestic flights had increased by an average of 2.7% since 1987. During the same period, the number of passengers on international flights rose by an average of 1.4%. Traffic associated with the Summer Olympic Games should substantially increase travel figures during 1988. During 1985, Korean Air transported 252,604 tons of cargo on both domestic and international routes. Because of projected growth in international air traffic to and from Korea, plans are being made to build a new international airport near Chongju, which is located approximately 100 kilometres south of Seoul.

Marine transportation. The Korean government manages its important marine transportation system through the Korean Maritime and Port Administration. This industry has been and will continue to be crucial Korea's development as a trading nation. In terms of overall tonnage, Korea has the fifteenth largest merchant fleet in the world. The vessels operated by Korean companies collectively totalled more than 7 million gross tons at the end of 1985. Korea has 24 first grade ports and 22 second grade ports. Substantial investments will be made to improve port facilities and operating efficiency in order to cope with increasing traffic demands and rising labour costs. By 1991, Korea's total marine cargo volume is expected to be approximately 250 million metric tons. The nation's maritime fleet will need to grow by 3 million gross tons to meet this demand.

Aerospace. Korea's fledgling aerospace industry is widely seen as ripe for spearheading the next stage of Korea's industrial development. A stronger aerospace industry, it is argued, would bring about many industrial and technological developments. It would also mean that Korea could reduce its dependence on the US for military equipment. Major emphasis will be placed on gaining technology through technical licenses and experience through producing parts and components.

Korean Air. The first company into the aerospace business was Korean Air which, in 1976, won a licence from McDonnell Douglas to assemble military helicopters. The company's aerospace division now employs 1,600 and has a projected turnover of W80 billion for 1988 (W40 billion in 1987). Its present activities include the assembly of F-5 fighters for Northrop, making wings for Boeing 747s and a maintenance contract for US military aircraft. The company is now planning the co-production of Sikorsky's UH-60 helicopters. The company has invested US$180 million to date, with a further US$120 million planned for the 1990s.

Samsung. In 1979, Samsung won a contract to overhaul engines for the US Air Force. Since then the company has won contracts from Pratt & Whitney and GM/Allison. In 1986 it began a US$350 million five-year plan aimed at boosting its aerospace capacity. Samsung has been selected by the government to take the lead in the drive to produce a fighter aircraft in association with a foreign manufacturer.

Daewoo. In 1984, Daewoo started making parts for General Dynamics F-16s and followed this in 1986 when it won a contract to supply wing parts for Boeing 747s. The company has so far invested US$60 million and is planning a further US$100 million.

Each of these companies is assured of heavy backing over the forthcoming years even though sustained profits are not expected for some time. The day when an aircraft designed and built in Korea is ready for domestic use and export is still some time away. However, this has not stopped the industry from achieving sustained growth. In the meantime, the aerospace companies have concentrated on diversifying its sub-contract base to include European manufacturers.

Telecommunications. Investment in the telecommunications sector lagged behind other sectors of the Korean economy until the 1980s. At the end of 1979, there were only 240,000 telephone subscribers in the entire country with a ratio of telephones to people of about 6:100. The task of improving the telecommunications system of the nation was assigned to the newly created Korea Telecommunications Authority in 1982. Also established at that time was the Data Communications Corporation of Korea.

As a result of increased investments in telecommunications made during the fifth EDP, the number of subscribers rose to 6,517,395 in 1985. This increased the telephone ratio to 16:100. The Ministry of Communications has established a target of 22 million total telephone lines and a telephone to citizen ratio of 40:100 (which is equal to levels in developed countries) to be met by the year 2010. The Korean Telecommunications Authority is moving toward establishing a nationwide digital switching and communications system. The nation's telecommunications system is also being upgraded to allow for increased data communications including simultaneous voice and non-voice transmissions.

Housing. Despite huge increases in housing construction in Korea since the 1960s, there continues to be a major shortage of suitable housing, particularly in urban areas. Continued migration to urban areas combined with the increasing population rate have made housing demand in Korea greatly outstrip supply. In 1980, the per capita living space in Korea was only 9.0 square metres. For this reason, both the fifth and sixth EDPs have placed an increasing emphasis on the construction of new apartments. During 1985 some 227,000 new housing units were constructed. At the end of December that year, approximately 14% of the nation's 6.27 million housing units were apartments. The percentage of Koreans living in apartments will continue to rise reflecting Korea's increasing urbanization.

Foods. Apart from the fillip that the Olympic Games will give, the domestic food industry looks set for a mature-to-declining future. Sugar demand is beset by global overproduction and the substitution of artificial sweeteners, flour is vulnerable to international market movements as it is entirely imported, and canned foods and dairy products are variable in quality and must be processed close to production centres. The brightest prospect is currently the instant noodle market but raw material costs are high compared to manufacturing. Given the low margins endemic in this industry and the trend toward imports, it is likely that new investment will be restrained.

Beverages. The consumption of alcohol in Korea reflects the rise in disposable income, sophistication, and western influence. Beer production overtook soju production in 1980 and rice wine sales have halved in eight years. Alcoholic beverages are expected to show a long-term growth of 5-7%, with above-average rates for high-priced, high-quality liquors such as whiskey and wine.

Over the last three decades, soft drink production has grown by 35% annually. Domestically produced soft drinks are exported to Japan, Malaysia and the Middle East, totalling W11,444 million in 1986. The 1988 Olympic Games should boost sales all out of proportion, but the market appears to have the capacity to sustain substantial future growth.

Paper and paper products. Korea is the 15th largest producer in the world, exporting US$140 million to Southeast Asia and the Middle East in 1986, and with a domestic market for paper container products expected to top W100 billion in 1988. Originally an import substitution industry in the 1950s, government assistance from the first EDP enabled it to meet domestic demand by 1973. Local industrial needs provided the additional impetus to tackle global demand so that by 1986 the facilities were operating at near capacity (96.4%). However, domestic pulp production accounts for less than half of the consumption and the industry is sensitive to worldwide price rises.

During 1987 Hyundai sold an estimated 400,000 cars abroad

VCRs and colour TVs are expected to be the export market leader with a forecast growth of 30% annually for the next five years

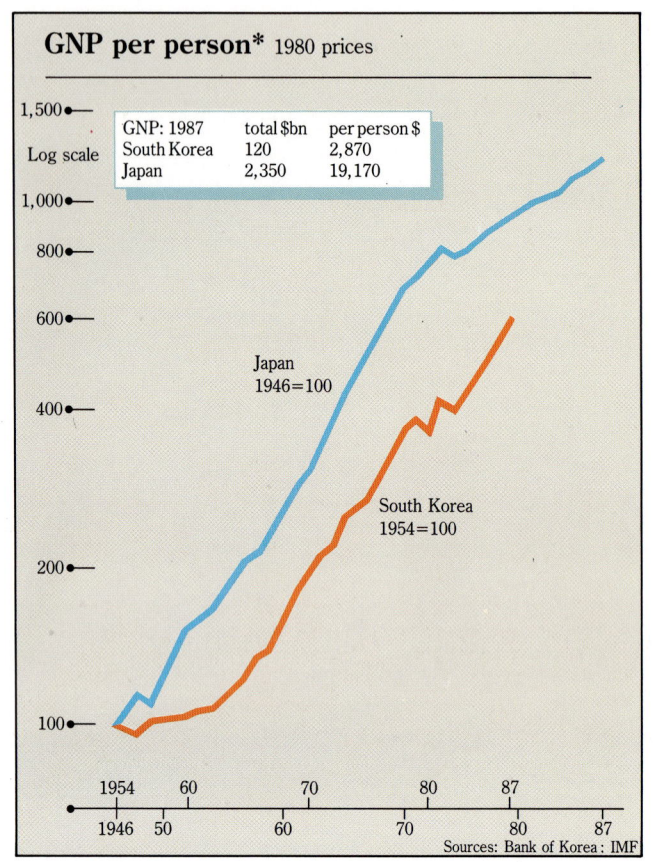

GNP per person* 1980 prices

GNP: 1987	total $bn	per person $
South Korea	120	2,870
Japan	2,350	19,170

Japan 1946=100

South Korea 1954=100

Sources: Bank of Korea; IMF

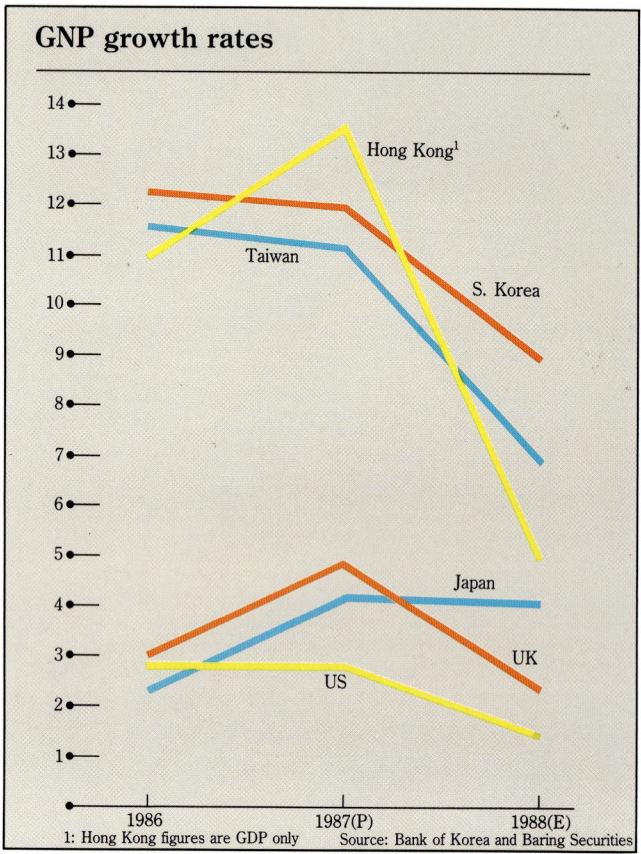

GNP growth rates

Hong Kong[1]

Taiwan

S. Korea

Japan

UK

US

1: Hong Kong figures are GDP only   Source: Bank of Korea and Baring Securities

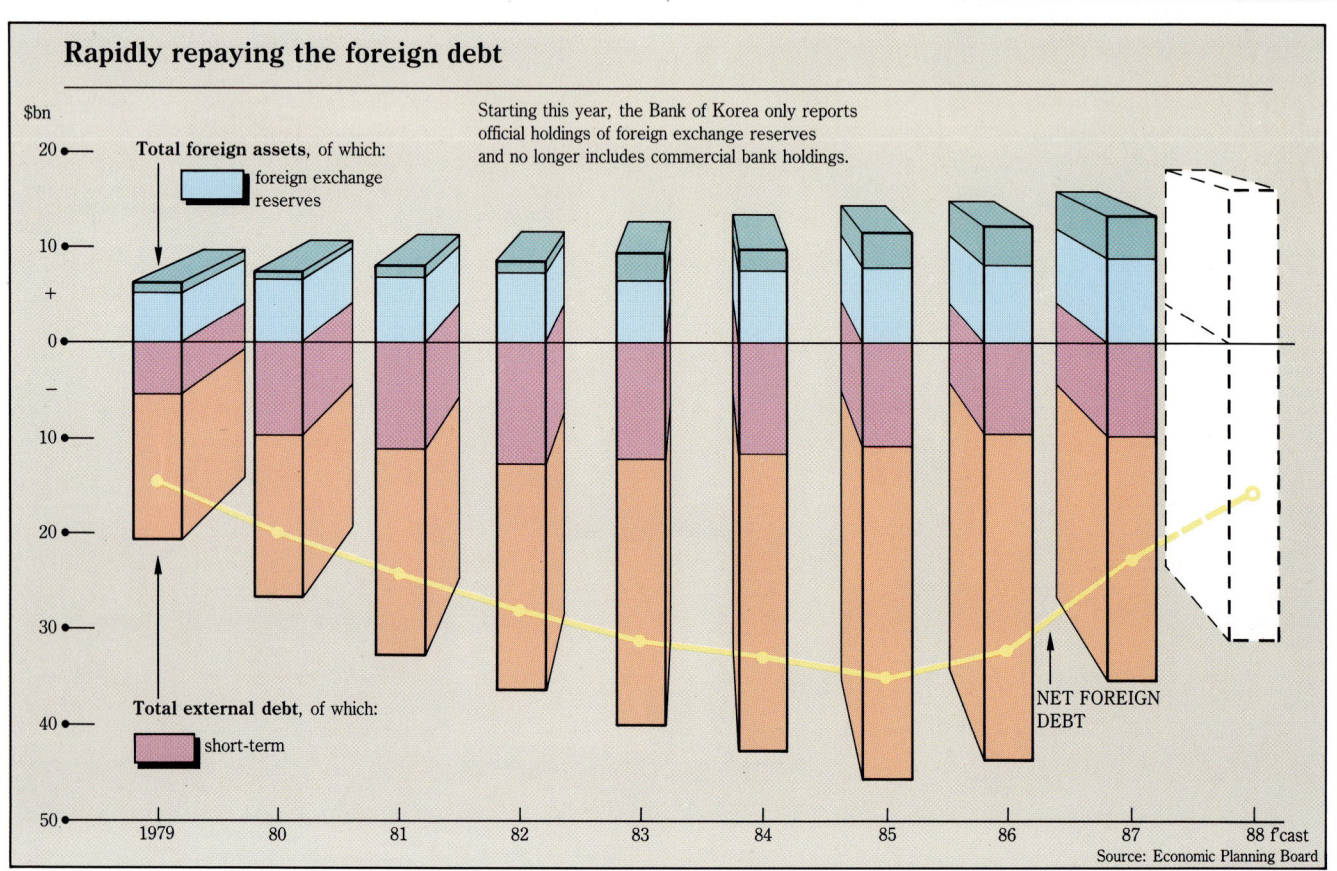

Rapidly repaying the foreign debt

$bn

Total foreign assets, of which: foreign exchange reserves

Starting this year, the Bank of Korea only reports official holdings of foreign exchange reserves and no longer includes commercial bank holdings.

Total external debt, of which: short-term

NET FOREIGN DEBT

Source: Economic Planning Board

48

The Five Year Economic Development Plans

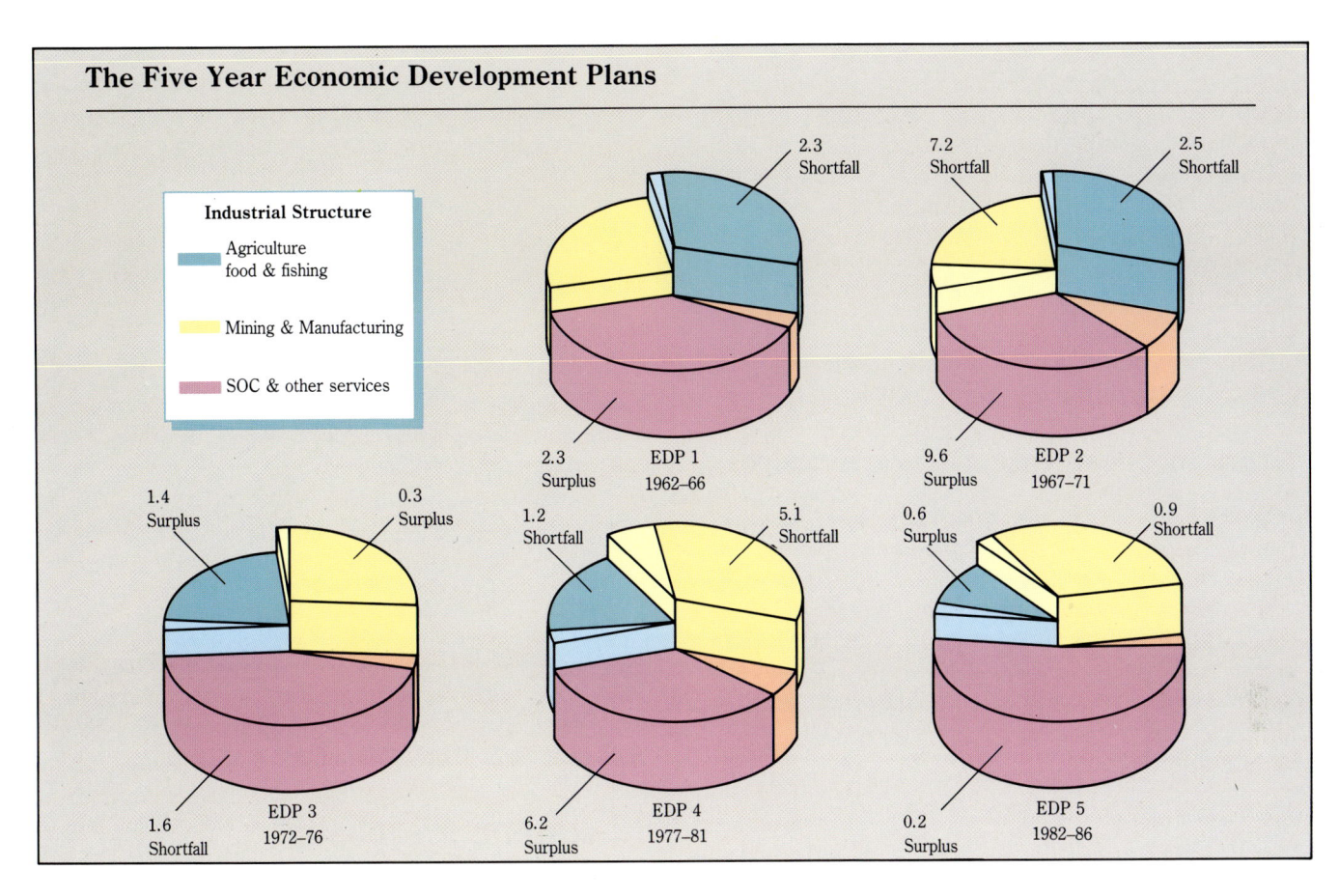

Industrial Structure
- Agriculture food & fishing
- Mining & Manufacturing
- SOC & other services

EDP 1 1962–66
- 2.3 Shortfall
- 2.3 Surplus

EDP 2 1967–71
- 7.2 Shortfall
- 2.5 Shortfall
- 9.6 Surplus

EDP 3 1972–76
- 1.4 Surplus
- 0.3 Surplus
- 1.6 Shortfall

EDP 4 1977–81
- 1.2 Shortfall
- 5.1 Shortfall
- 6.2 Surplus

EDP 5 1982–86
- 0.6 Surplus
- 0.9 Shortfall
- 0.2 Surplus

Approximately 60% of the land area of Korea consists of uncultivated and forested slopes

Pharmaceuticals. Until the Korean War the pharmaceutical industry was a small-scale operation, producing pills from Chinese herbs, digestive drugs, boric acid ointment, and simple medications. The considerable restructuring brought about by the war made possible the production of analgesics and sulfa drugs, antibiotics, anti-tuberculosis medications and vitamins. During 1965-75, the industry was able to satisfy 95% of local demand. In 1986 the value of production totalled W1.9 trillion (US$2.2 billion), exporting US$1.3 billion. The Korean Pharmaceutical Industry Association predicts a compound growth rate of 10% during this decade.

Plans for economic diversification

In order to move Korean industry toward the production of goods with a greater value-added component, the government has targeted ten strategic items for development. Substantial investments in research and development as well as production facilities are planned in order to raise export levels of these products. Approximately 40 Korean companies involved in exporting these products will be given financial and other assistance. The products chosen as promising export items are:

Non-athletic footwear. Efforts will be made to increase quality, diversify product lines and establish sales of products bearing Korean trademarks. Companies will work to reduce dependence upon athletic shoes by improving local design skills.

Microwave ovens. Sales of Korean microwave ovens are approaching a 40% share of the US market. Sales are also growing in Europe. Continuing to centralize the production of parts and components is planned in order to increase profitability without increasing protectionism. More reliance will be placed on Korean brands rather than producing for foreign private labels.

Pianos. Korean companies such as Samick and Youngchang are becoming well known on world markets. Production facilities for pianos and other musical instruments will be expanded significantly.

Furs. Korea has become a leader in the production of moderately priced furs. Companies like Jindo and Taelim have been very successful in manufacturing and selling fur garments in world markets. Pelts are imported so their competitive advantage comes from lower labour costs. Quality should improve in years to come.

Fishing equipment. Korea is the world's largest exporter of these products. Quality gains will increase exports of fishing rods and reels. Localization of blank rods will be accelerated.

Stereo equipment. Growth in sales of stereo equipment should be assisted by the increase in the value of the Japanese yen. Attempts will be made to establish Korean-owned brand names.

Glasses frames. Korea's small and medium companies have been successful in capturing a niche in the low end of this market. Efforts will be made to increase both quality and average export prices.

Ceramic tableware. Korea is hoping to capitalize on its centuries-old skills in making and designing ceramics. Korea has a good supply of local clays.

Leather handbags. Improvements in product quality are also due for this product. More funds will be invested in improved design.

Leather products. Efforts are being concentrated on improving tanning and dyeing skills. Korea has long been a leader in the production of leather-topped athletic shoes. The production of dress shoes has not met expectations.

Economic forecasts

Near-term economic prospects

Like Japan, Korea has made fools of economic forecasters who have doubted the country's ability to adjust to changing world economic conditions. The economy of Korea has survived two oil shocks, the Middle East construction bust, significant labour and political unrest and increasing foreign protectionism. Nothing in its recent track record suggests that Korean economy is due for a fall in the near future.

Korea must travel through uncharted waters and will face many challenges in years to come. The nation's economic planners realize that Korea must produce higher value-added goods and move from being a technology assimilator to being a technology innovator. Despite significant increases in domestic demand, exports will continue to be relied upon to increase economic growth. While domestic consumption will increase, per capita income has not reached a point where industry can concentrate heavily on domestic markets. Furthermore, Korea's relatively limited population can never be expected to be large enough for manufacturers of sophisticated products to recover redevelopment costs if they sell only domestically.

Set out below are discussions of major sectors and areas of concern relating to the Korean economy.

Inflation. Inflation has been a major problem during nearly all of Korea's economic development since the 1960s. As recently as the late 1970s Korea experienced double-figure price increases. The Seoul Consumer Price Index increased at an average rate of 15.2% per annum from 1960 to 1980. The combination of the *three blessings* and an effective monetary policy has recently achieved impressive price stabilization. However, the battle with inflation is not over yet as the consumer price level increased from 2.3% per annum in 1986 to 5.8% in 1987. Much of the increases are attributable to the *cost-push* effects of rising labour costs which for several years have been masked by falling oil prices.

To combat increased inflation, the Korean government announced in early 1988 that it will pursue a tight monetary policy, reduce tariffs on imported raw materials, discourage real estate speculation, restrain increases in government spending, lower government controlled fuel prices and stabilize the supply of imported products. President Roh has stated publicly that there will be an all out effort to keep the growth of the money supply below 20%. Price freezes on some basic commodities can be expected and in addition, certain indirect taxes may be reduced or eliminated. The government has stated that it will do everything possible to keep inflation below 5% in 1988, but will pay particular attention to the prices of basic commodities needed by families. "Growth through stability" will continue to be a major government policy goal. This will be very difficult given the need to pay for social welfare projects, the spending growth associated with the Olympic Games and the flow of foreign

One major component of the government's monetary policy will be a restriction on lending by Korean banks. In general, loans for general commercial purposes are not likely to be approved unless they are crucial to the country's economy. It has been reported that the government no longer allows loans to large corporations even if they involve exports. As an indication of the government's policy, it is noteworthy that total lending by the seven city banks increased by only 2.9% during 1987.

Wage rate increases. The wage rate increases given to Korean workers will do much to determine the vitality of the Korean economy. The labour component of manufactured goods averages approximately 50% of value. Wage increases during 1987 were substantially offset by an 11% growth in productivity. Most Korean economists predict that wages will increase by approximately 15-20% in 1988, though the wage increase of over 20% given by Daewoo Shipbuilding in early 1988 may set a precedent that other Korean companies might follow.

Unemployment. The Economic Planning Board predicts that unemployment will range between 3.7–3.8% during 1988. This represents approximately 550,000 persons out of a total workforce of 15.9 million. It is expected that 370,000 new workers will enter the labour force in Korea during 1988. Of these new workers, 10,000 will not be able to find employment.

Exports and imports. Korea's economy is particularly dependent upon international trade. Not only is Korea dependent upon imports to fuel its economy, it is vulnerable to reductions in the demand for its exports. Despite increased protectionism in foreign markets, the Korea Development Institute predicts that exports from Korea will increase to US$52.2 billion in 1988 and imports will rise to US$47.3 billion. To understand the progress which Korea has made, it is necessary to consider that in 1962 exports had a value of only US$55 million and imports a value of US$422 million. The progress of Korea as a trading nation since 1962 is nothing short of astounding.

Because Korea is a resource-poor nation, imports have and will continue to be dominated by raw materials. In 1985, Korea imported US$5.6 billion of petroleum and US$11.8 billion of other raw materials. This represents over 50% of 1985 total imports of US$31.1 billion. Korea is also a significant importer of grains, bringing in US$6.7 billion in grains from 1980 to 1985. The annual rate of growth of imports (in real terms) averaged approximately 20% between 1965 and 1981.

In the industrial category, Korea's major import items are transportation equipment, machinery, and electronic components. In terms of import dependence, Japan is clearly the most important supplier of Korea's imports. For example, Japan supplied more than 40% of Korea's capital goods imports in 1986. Similarly, over 40% of parts and machinery are sourced from Japan. Over 80% of Korea's trade deficit was with Japan during the 1965-86 period. Because of this import dependence upon Japan, the rise in the value of the yen in recent years has seriously damaged Korea's ability to compete abroad.

In order to protect domestic industries and to conserve foreign exchange, Korea has created a variety of import restrictions. These import curbs take the form of quantitative limitations, tariffs, standards, import prohibitions, government guidance and consumer campaigns against imports. These restrictions are described more fully later in this book.

Over the past decade, Korea's export competitiveness has been under constant pressure due to the rising value of the won, increased labour costs and increased competition from developing nations. It is this pressure that has changed the nature of Korea's exports. The nine largest export categories for Korea in 1986 were textiles, electronics, iron and steel products, footwear, ships, automobiles, marine goods, general machinery and electrical equipment. The automobile category should continue its rise as investments in this area have been substantial in recent years. Electronics should become the leading export category during 1988 or 1989.

The US has been, and will continue to be for some time, Korea's largest export market. The US market accounts for approximately 40% of Korea's exports in 1986. Japan was responsible for 15.5% and the European Community

12.4% that year. Despite considerable efforts to diversify markets, sales by Korean companies to Latin America, Africa, the Middle East and other developing nations have actually fallen in relative terms.

The increase in exports shows no sign of slowing. Korea's exports rose 36.1% in 1987 from record 1986 levels to US$47.28 billion. Growth in exports during 1987 was broadly based but was led by the technology intensive sectors. Growth rates in some key sectors of the economy during 1987 were:

electronics	54%
automobiles	106.6%
textiles	33.6%
footwear	33.6%

Imports into Korea have risen but not as quickly as exports. During 1987, imports had a value of US$41 billion, approximately a 30% increase from 1986. The rate of growth for consumer goods imports during 1987 was 27%, capital goods 29% and raw materials 31%. Imports during 1988 have been projected to increase to US$49.6 billion by the Ministry of Trade and Industry.

Because of Korea's export dependence, the next two years will be crucial in determining whether the growth in Korea's economy can be sustained. Reduced demand overseas and increased protectionism in foreign markets could cause considerable problems for Korea's economy. The Korea Institute for Economics and Technology (KIET), a government-funded think tank, has projected that growth in the world economy will slow to 2.2% in 1988. This is less than the 2.6% growth recorded in 1987 and the 2.9% growth in 1986. Increasing domestic wage levels mean that Korea must move away from labour intensive activities to more technology-intensive high value-added products. In addition, Korea must rely more heavily on domestic markets and export markets outside the US and the European Community. The government's watchwords for this shift are "balanced economic expansion" rather than the previous "export first" slogan.

The economic forecast for 1988 is clouded by the general uncertainty in the world economy. Korea is perhaps uniquely dependent upon the economies of other countries because of its high export profile. Exports currently comprise approximately 40% of Korea's GNP. Huge facilities investments have been made, based upon a company's ability to export most of its output. The Ministry of Trade and Industry has recently revised its projections and predicted that Korea's exports should grow to 13% from 1987 levels to approximately US$52 billion, despite the expected appreciation of the won against the US dollar and the increasing import restrictions imposed by advanced countries. These forecasts should be viewed with some caution since the Korean government has recently learned to conservatively estimate the projected growth of exports and the economy in order to reduce protectionism in foreign markets.

Balance of payments. Korea had a total current account surplus of US$9.78 billion in 1987. This amount is more than double the surplus of US$4.62 billion recorded in 1986. The probability is that Korea's current account surplus will fall only slightly during 1988 and 1989. The overall balance of payments was a surplus of US$5.26 billion in 1987, which is over 300% more than the US$1.7 billion surplus in 1986.

The surplus is being used in large part to repay foreign debt (much of which has high interest rates or unfavourable terms). This debt retirement activity will reduce the nation's total interest burden and thus increase the total surplus in 1988 and future years. The amount of interest paid on foreign debt fell from US$3.7 billion in 1986 to US$3 billion in 1987.

Improved tourism income (much of which will be associated with the Olympics) will also increase invisible trade surpluses during 1988. The Korea Development Institute believes that the current account surplus should, as a policy matter, not be greater than 5% of GNP. The level of current account surplus against GNP for 1987 was 8.3%. A 1988 current account surplus of less than US$7 billion would meet the 5% standard. To meet this objective, the current account surplus would need to be reduced by one-third from 1987 level. The KDI has indicated that the gov-

Ministry of Trade and Industry import projections for 1988

By type	1987	1988 (in US$ billion, %)	
		Projection	Change (%)
Total imports	40.5	49.6	22.2
Raw materials	22.1	27.2	23.1
Capital goods	14.4	17.6	22.2
Consumer goods	4.0	4.8	20.0
Export related	17.9	22.1	23.5
Domestic consumption	22.6	27.5	21.7

By industry	1987	1988 (in US$ million, %)	
		Projection	Change (%)
Total imports	40,574	49,600	22.2
Agro/fisheries/forestry	2,990	3,415	14.2
Minerals	6,644	7,855	18.2
Industrial chemicals	8,290	10,193	23.0
Textiles	2,590	3,238	25.0
Steel & metals	3,985	4,790	20.2
Machinery	15,545	19,473	25.3
Sundry goods	530	635	19.8

ernment should reduce the nation's current account surplus to approximately 2% of GNP by the 1990s.

To reduce protectionist sentiment, the Korean government is moving to lower the growth of the trade surplus with the US which rose to US$9.55 billion in 1987 (a 57% rise from US$7.34 billion in 1986), while simultaneously reducing Korea's trade deficit with Japan. Government policy makers have announced their intent to reduce the 1988 trade surplus with the US to US$6 billion and the current account surplus to US$3 billion.

The Korean government has also announced its intention to lower import tariffs, allow the value of the won to appreciate, increase overseas investments, reform customs administration procedures for imports, and make it easier for Koreans to travel and study overseas. To this end, Korea has embarked on a campaign to produce more parts and components domestically since to a large extent these parts come from Japan. This localization process will be a huge task given the high level of dependence upon Japan as a supplier of parts, technology and capital. The government's efforts to reduce the trade deficit with Japan began to bear fruit in 1987 as it came down to US$5.22 billion, a reduction of US$220 million from 1986. During 1987, imports from Japan fell by 25.6%. This is a significant accomplishment given the increased level of exports from Korea.

The Korean government has adopted a well-publicized "buy American" campaign as part of its effort to reduce the Korea/US trade imbalance. A major component of this campaign is the government sponsorship given to *buying missions* sent to the US. To date, purchases made on such buying missions have been limited to *big ticket* capital goods and raw materials (e.g., airplanes, corn, wheat). Korean Air has announced large orders for Boeing commercial aircraft in 1988 as part of the drive to import from the US. Fewer purchases from the US have involved machinery and parts and components. This is the area with the greatest growth potential for Korea/US trade as western companies have generally done poorly when competing with Japanese companies to supply parts and components.

Underground economy. The Korea Economic Research Institute (KERI) issued a report in 1987 estimating that Korea's underground economy may have been as large as 30% of GNP in 1986. This would put the size of the underground economy at US$30 billion that year. Similar ratios against GNP for other nations are much smaller (for example, 13% in the US in 1981 and 25% in Taiwan in 1979). This underground economy principally flourishes in urban rather than rural areas. If the underground economy is taken into account, urban per capita income in Korea may exceed US$5,000 according to some experts. The principal unreported activities include kerb market lending, securities and real estate speculation, smuggling, tax evasion, corruption, and black marketing.

The Office of National Tax Administration has estimated that the total amount of tax evasion during 1984 was W2.7 trillion or 21.4% of taxes collected. Informed observers believe that this figure is low. Korean businessmen are reputedly poor record keepers, particularly those who have not incorporated. Taxes paid by unincorporated businesses tend to be assessed on the basis of an industry average rather than actual revenues. One major area in which taxes are under-reported involves the entertainment industry. Nearly 900,000 Koreans earn tip incomes and as many as 100,000 in the industry are believed to live solely from tips. Little of this income is reported.

Savings, investment and consumption rates. The savings rate rose to 36% of GNP in 1987 which is significantly higher than the country's investment requirements (which are estimated to be around 31% of GNP). There is no reason to expect that the savings rate will be any less in 1988. These savings can be used to fuel Korea's economic growth and to further reduce the amount of foreign borrowed capital. Fixed investment is predicted to increase by 10–11% in 1988 largely due to the emphasis on housing construction by Korea.

Foreign investment. Foreign investment should remain strong in 1988 and 1989 but will be reduced from 1987 levels. Investments from abroad in the hotel and automobile industries will continue but will be smaller in amount. In short, a larger number of smaller investments can be expected, producing a slightly reduced net level of investment. To avoid the flow of foreign money to Korea from investors hoping to take advantage of currency appreciation, the government will require foreign investors to precisely disclose the manner in which funds will be used. The government will also reduce tax benefits available to foreign investors and will be more selective in granting foreign investment approvals.

Protectionism and trade negotiations. The decision made in 1987 by the US to revoke Korea's GSP (Generalized System of Preferences) status completely from 1989, instead of merely removing a few Korean products from GSP status each year is an indication of a *get-tough* trade policy. Importers of Korean goods into the US will pay an average of 5% more in duty due to the loss of GSP benefits. Section 301 lawsuits have been filed by US beef, wine and tobacco producers, and further action on trade disputes between the US and Korea is likely. To offset this, recent trade concessions have been made in the cigarette, beef and orange juice sectors, but early 1988 trade figures show an increase in Korea's trade surplus with the US of US$400 million from the same period in 1987. Protectionism is also likely to increase in the European market as a result of

Poster urging Korean people to exercise their right to vote

Korea's trade surplus with Europe, which jumped 50% to nearly US$3 billion in 1987.

The Korean government will have a difficult time in granting trade concessions to the US or the European Community in light of the fact that the ruling DJP still needs to consolidate its domestic political position. The Roh government can be expected to plead strenuously for time in an effort to further the democratization of Korea. It must be emphasized that the Korean government does not have much room to manoeuver in granting trade concessions. Recent Korean offers to lower imported cigarette prices or allow sales of US beef in tourist hotels are not likely to affect the trade deficit to any significant degree. Trade concessions which take a real bite out of the trade surplus will be political dynamite. Opposition politicians will use any trade concessions to bolster their popularity as the electorate views such concessions very unfavourably.

Currency depreciation. Tremendous pressure will be applied by the US and European governments for a revaluation of the won. While a one-off revaluation of the won is unlikely, a gradual depreciation by as much as 15% against the US dollar is likely in 1988, although a 20% appreciation of the won against the US dollar is not unthinkable. Appreciation of the won against foreign currencies should gradually diminish until it stabilizes by the mid-1990s. The Korean government is likely to adjust the multi-currency basket used to calculate exchange rates to reflect a relatively greater increase of the value of the won in relation to the Western European currencies. The Korean government has more than sufficient foreign exchange reserves to protect the currency's value if it should become necessary.

Korean economic forecasters are not in agreement on all issues concerning the effect on the domestic economy of a depreciation in the value of the won. Since the depreciation should lower real GNP, increases in imports due to their more attractive price should to some degree be offset by reduced demand. The Korea Development Institute and the Bank of Korea have released studies which predict that a 10% depreciation of the value of the won will reduce GNP by approximately 2%. The studies disagree on the relative effect this depreciation would have on the trade balance. The Bank of Korea predicts a US$0.8 billion marginal decrease while the Korea Development Institute estimates the marginal reduction to be US$1.6 billion.

Summary of Economic Development Plans

	First EDP[1]	Second EDP	Third EDP	Fourth EDP	Fifth EDP
Principal Objectives	1. Correction of socio-economic vicious circle of previous years. 2. Establishing the foundations of a self-reliant Economy.	1. Modernization of Industrial structure. 2. Promotion of self-reliant economy.	1. Coordination of growth, stabilization, and equity. 2. Achievement of self-reliant economy. 3. Maximization of land cultivation and balanced regional development.	1. Achievement of self-generating growth. 2. Promotion of equity through social development. 3. Technical innovation and efficiency improvements.	1. Stabilization of prices. 2. Raising of productivity. 3. Promotion of social development. 4. Elimination of distortions in the economy and promotion of rational development.
Period	1962–66	1967–71	1972–76	1977–81	1982–86
Economic Growth Rate[2]	7.1 (8.5)	7.0 (9.7)	8.6 (10.1)	9.2 (5.5)	7.56 (8.729)
Major Policies	1. Securing energy supply sources. 2. Correction of structural unbalance in economy. 3. Expansion of key industries and social overhead capital. 4. Effective use of idle resources. 5. Improvement of balance of payments position. 6. Technical development.	1. Self-sufficiency of grains and development of forestry and fishery. 2. Formation of foundation for industrialization. 3. Improvement of balance of payments position. 4. Raising employment ratio and restriction of population increase. 5. Raising farmer's income. 6. Improvement of technology and productivity.	1. Self-sufficiency in food staples. 2. Improvement of living environment in rural and fishing villages. 3. Improvement of balance of payments position. 4. Improvement of industrial structure. 5. Development of technology and human resources. 6. Expansion of social overhead capital. 7. Effective use of resources. 8. Improvement of welfare.	1. Securing domestic investment resources. 2. Achievement of balance in B.O.P. 3. Raising international competitiveness. 4. Expansion of employment opportunities and manpower development. 5. Expansion of Saemaul campaigns. 6. Improvement of living conditions. 7. Expansion of Investment in science and technology. 8. Improvement of economic management and system.	1. Curbing inflation. 2. Recovery of competitive power in heavy industry. 3. Consolidation of agricultural policy. 4. Improvement of financial system. 5. Overcoming energy constraints. 6. Adjusting government function and rationalization of fiscal management. 7. Shift to competitive system and promotion of open economy policy. 8. Development of education and manpower and promotion of science and technology. 9. Establishing new relationships between labour and management. 10. Expansion of social development.

Source: Economic Planning Board.
Note: 1) Economic Development Plan.
2) Figures in parentheses denote the actual growth rate.

Chapter VII
Economic projections

In trying to assess the development of the Korean economy in the near- and medium-term, it is useful to examine the sixth EDP adopted by the Korean government. The government's major objectives in formulating the sixth EDP were to increase economic stability and to improve economic efficiency. Much of this plan involves increasing competition by loosening government controls on the economy and by internationalizing markets. Industries will be encouraged to increase localization of parts and components and to diversify export markets. Other important components of the plan are increases in professional management, increased separation of ownership and management and improved technical innovation. Major efforts will be made to improve the distribution of wealth, increase housing supplies and decrease business concentration. Price stability and low unemployment will receive considerable policy emphasis.

Projections for the Sixth Economic Development Plan

Category	Unit	1987	1988	1989	1990	1991	Average ('87-91)
Population	1,000 persons	42,094	42,593	43,099	43,601	44,094	
GNP growth	%	7.7	7.5	7.0	7.0	7.0	7.2
GNP	$bn	103.9	117.7	131.9	147.8	166.0	
Per capita GNP	$	2,474	2,767	3,065	3,396	3,800	
Unemployment	%	3.7	3.6	3.7	3.7	3.7	3.7
Balance of payments							
current A/C	$bn	2.3	2.7	3.1	3.6	4.0	
trade A/C	$bn	3.0	3.5	3.9	4.4	4.8	
exports	$bn	35.6	39.8	44.2	49.1	54.4	
(growth)	%	(12.3)	(11.8)	(11.1)	(11.1)	(10.8)	(11.4)
imports	$bn	32.6	36.3	40.3	44.7	49.6	
(growth)	%	(11.6)	(11.3)	(11.0)	(10.9)	(11.0)	(11.2)
Inv. trade	$bn	−1.4	−1.5	−1.5	−1.5	−1.5	
(net trans.)	$bn	0.7	0.7	0.7	0.7	0.7	
Outstanding							
ext. debts	$bn	47.9	48.2	48.1	47.3	46.1	
(net ext. debts)	$bn	(34.3)	(32.3)	(30.1)	(27.1)	(23.7)	
Prices							
wholesale	%	2.0	2.0	2.0	2.0	2.0	2.0
consumer	%	3.0	3.0	3.0	3.0	3.0	3.0

Source: Economic Planning Board

Despite political uncertainty and labour unrest during 1987, Korea's economy grew at an adjusted rate of 12.2% according to the Bank of Korea. This growth rate was only slightly lower than the 12.5% growth which occurred in 1986. Thus, economic growth significantly exceeded the projections of the sixth EDP in both 1987 and 1988. As has been the case since the early 1980s, increased exports were the primary reason for the growth of the economy. To reflect this faster than expected growth, the Economic Planning Board in early 1988 increased its estimate of average annual growth during the plan from 7.2% to 8.3%.

Comparison of 1988 projections by government and affiliated organizations

	KDI	BOK	EPB	KIET
Growth rate (% GNP)	8.5	8–8.5	8	8.3
Trade balance (US$ billion)	3.5	5–5.5	4.5	6.6
Current account balance (US$ billion)	6.5	7.5–8	6	– –
Unemployment (%)	3.7	3.6	3.7	– –
Wholesale prices (%)	3.5	4	2–3	– –
Retail prices (%)	5	6	4–5	5.5
Per capita GNP (US$)	– –	– –	3,300	– –

However, the figures for the first two months of 1988 were so strong that the EPB decided to revise its projections.

Revised Economic Planning Board projections

GNP (US$ billion)	145.0
Per capita GNP (US$)	3,450.0
Domestic savings ratio (%)	35.5
Total investment ratio (%)	32.0
Current account (US$ billion)	7.0
trade balance	5.0
exports	55.0
imports	52.0
invisible trade balance	2.0
Gross external debt	31.0
Overseas assets	15.5
Net external debt	15.5

If the above projections are met, the economy will grow at 8%, assuming an 8% increase in total consumption, 9.5% increase in fixed investments, and a 16.3% export increase. Consumer prices are projected to increase by 5% and R&D to 2.4% of GNP.

The economic predictions of Korean forecasting organizations for 1988 vary significantly in many cases. For example, the Korea Institute for Economics and Technology (KIET) has predicted that the trade surplus in 1988 will be US$8 billion in 1988 while the KFTA predicts that the surplus will be only US$3.5 billion. In general, the forecasts of these and other organizations affiliated with the government are fairly conservative and reflect the government's desire to deflect criticism originating from Korea's trading partners and to reduce the amount of protectionist legislation enacted in other countries.

Independent forecasts for the three years to 1990

	1988	1989	1990
Economic growth (% GNP)	12	10	9.5
Wholesale prices (%)	4.5	3.5	3.0
Consumer prices (%)	7.5	5.5	4.5
Trade balance (US$bn)	7	6.5	6
Current account balance (US$bn)	9.0	8.5	8
Trade surplus within US (US$bn)	8	6.5	6
Prime rate (% at year end)	10	9.5	9.5
Exchange rate (W/US$ at year end)	670	620	570
Stock market (CSPI)	760	850	1000
Unemployment	3.5	3.4	3.3

Medium- and long-term prospects

The KDI predicted in 1987 that the Korean GNP will grow at an average rate of 7.2% per annum until the year 2000. Population increases should be 1.8% from 1987–1991 and 1% until 2000. The projections show that the labour force will grow at a rate of 3% for much of this period, reducing the number of dependants each worker must support but increasing the need for new jobs. The population of Korea in the year 2000 should be approximately 50 million.

Korean Development Institute growth projections

	1991	2000
GNP per capita (current prices in US dollars)	4,000	6,900
GNP deflator (1980=100: in %)	3.5	3.9
Exports (in billion current US dollars)	56	173
Imports (in billion current US dollars)	53	168
Current account balance (in billion US dollars)	5	5
External debt (in billion US dollars)	33	(surplus)
Unemployment rate (in %)	3.7	– –
Savings rate (in %)	33.5	36.5
Investment rate (in %)	31.3	35.3

This gives Korea a level of per capita income equal to Italy's 1987 level.

The KDI, in a report entitled "Changes in Future Industrial Structure and Policy Measures" issued in January 1988, provided revised and extended medium-and long-term forecasts for the Korean economy. The change was necessary as growth projections in the initial study by the KDI were far too conservative. For example, exports may exceed projected 1991 levels as early as 1989. The new study predicts the following percentage annual growth rates for the Korean economy:

1986–1991	8.6%
1992–2000	6.8%
2001–2010	5.8%

The projected growth places GNP at US$271.6 trillion in 2010. This would mean that GNP in Korea (in 1985 constant price terms) would increase to US$10,000 per capita by 2010, putting Korea at the level of today's advanced countries. The study predicts that the share of employment held by agriculture and fisheries will fall from nearly 25% in the mid-1980s to 9% in 2010. The KDI predicts that Korea will have a cumulative current account surplus of US$126 billion in 2010.

The government announced in January 1988 that Korea plans to join the Organization for Economic Cooperation and Development (OECD) in 1992. At that time, the government projection is that per capita GNP will be in excess of US$5,000. To be ready to comply with OECD requirements, the Korean government must, well before that date, proceed with trade and foreign exchange liberalization. In addition to joining the OECD, Korea expects to be in a position to join the General Agreement on Tariffs and Trade (GATT) as well as become a member of the International Monetary Fund.

Many analysts have attempted to formulate a prediction of Korea's long-term position in the world economy. Although an over-simplification, one plausible view holds that Korea will remain a medium technology industrial power whose strengths lie in capital-intensive, high volume and low margin manufacturing. Developments in research will primarily relate to improving and refining technology developed in other industrial nations. Korea's high savings rate and education levels and the impressive degree of business concentration make this approach both necessary and desirable.

The revised long-term projections of the KDI are probably too conservative. Korea's strategic location on the Pacific Rim should accelerate economic development well beyond the government's forecasts. Korea should continue to expand its share of world trade despite protectionism in other markets. Simply put, Korea has made a wide range of economic investments which are likely to pay off in the decades to come. Only a major world-wide recession would prevent Korea from becoming a world economic power by 2020.

Recent economic developments

Korea first began to emerge from its self-imposed international commercial seclusion when it was forced by Japan to sign the Kanghwa Treaty in 1876, thus opening the nation's ports to international commerce. The economic backwardness of Korea at the time of this relatively late opening to international trade was unrivalled in Asia. Much of the lack of economic development in Korea at that time can be traced to failures in government. The power of the Korean king had been weakened by centuries of infighting between members of the royal family, court ministers and numerous factions within the upper (*yangban*) class. The government was able to build few roads because of the country's extremely mountainous terrain and because of the restricted royal treasury. Coastal trade between Korean ports was negligible owing to government imposed restrictions on maritime trade and low standards of vessel construction and maintenance.

Even among Koreans, commerce was discouraged because of the very low social status given to merchants under Confucian teachings. The Korean interpretation of Confucianism assigned merchants and other businessmen a social status near the bottom of the hierarchy, just above felons, prostitutes and slaves. Korean society was therefore dominated by scholars, government officials and upper class landholding families who had no desire to loosen their grip on society. As a result, Korea was almost entirely agrarian and had no strong merchant class at the time of the opening to international trade. Commerce in Korea was also handicapped by the lack of a viable domestic currency. Very few goods were produced outside the home and few factories or workshops were in existence. To make matters worse, the lack of foreign contact put Korea's scholars far behind their foreign counterparts in the sciences. For this reason, Korea had considerable difficulty in assimilating foreign technology even after international contacts began.

During the years between the signing of the 1876 Kanghwa treaty and full colonial occupation by Japan in 1910, trade between Korea and the outside world was negligible. While Japan did establish a few banks and railways as well as take some steps to improve agricultural methods, little real economic progress was made. The Japanese effort to develop agricultural land in Korea was conducted through an entity known as the Oriental Development Company. This entity would later seize huge tracts of land from Koreans through confiscation.

When formal annexation by Japan was completed in 1910, the occupation government quickly structured the Korean economy to serve the needs of the Japanese economy to the greatest extent possible. Initially, the occupation government moved to increase agricultural output to meet Japan's increasing need for food grains. Rice shortages were common each spring in Korea because too much grain was shipped to Japan. This emphasis on agriculture continued until 1920 when Japanese light industry began to establish manufacturing facilities in Korea to take advantage of the nation's low wages and cheap raw materials (which are located in what is now north Korea). Beginning in the 1930s, the Japanese occupation government began to structure the Korean economy to serve its military needs. For this reason, some heavy industry was developed in the northern half of the peninsula to take advantage of mineral deposits and cheap hydroelectric power. Many Koreans were forced to move to Japan to serve as cheap labour during this period.

Japan made little attempt to industrialize or train the Korean people during the colonial occupation. While infrastructure in Korea was marginally improved by the Japanese, the economy remained largely agrarian and subservient to the needs of Japan. Korea's resources were exploited during this period without concern for their proper long-term management. In particular, Korea's timber was ruthlessly stripped. In addition, the high taxes imposed by the Japanese occupation government left little capital for investment in the Korean economy.

When Japanese occupation ended in August 1945, and the US Military Government assumed control of governing Korea during a transition period, the major economic problem to be addressed was restructuring the Korean economy to reflect the separation from Japan. Post-war Korea also needed a radical increase in the number of skilled workers and engineers, since over 90% of businesses and 80% of skilled labour and engineers were Japanese. Manufacturing distribution was also unbalanced as nearly all heavy industry, including chemical plant and electric power generation capacity, was located in the territory occupied by the communists. As Japanese companies and technical personnel withdrew from Korea, production dropped markedly. To make matters worse, food was in continual short supply due to the huge influx of refugees from north Korea, wartime social dislocation and poor agricultural practices.

The Korean War, which began in 1950 and continued for three devastating years, brought the nation's economy to a near standstill. Nearly all of the nation's food and all finished goods were imported from abroad as in-kind foreign aid during and immediately after the war.

The ending of the war did not provide immediate relief. The Korean economy grew only at an annual rate of 4.4% from 1954 to 1961. Much of this growth merely involved the reconstruction of destroyed economic capacity with foreign aid and since the growth in the economy barely matched income levels, there was no real improvement in the standard of living. As the new decade began, Korea's citizens had little hope for the future and began to agitate for change.

The five-year economic development plans

In response to the widespread economic dissatisfaction when the Park government seized power in 1961, a decision was made to adopt a series of five-year economic development plans. When viewed in their entirety and individually, these plans are a remarkable success story.

The First Five-Year Economic Development Plan 1962–66

The plan was drafted by the newly-established Economic Planning Board largely on the basis of the military government's "comprehensive economic reconstruction plan." Because it was too optimistic, the first EDP had to be significantly revised in 1964. Its primary purpose was to "build an industrial base principally through increased energy production." The slogan for the plan was "increase production, exports and construction." Other goals of the first EDP included improving farm income by increasing agricultural production and improving infrastructure by building harbours and railroad lines as well as cement, fertilizer and refinery facilities. However, the most important component of this plan was the decision to improve the nation's economy by promoting exports. This export drive was not without its critics in a nation with such a long history of political and commercial isolationism. To promote exports from Korea, a wide range of financial subsidies and tax benefits were granted to exporters by the government. Other measures taken to stimulate exports included a large devaluation of the won and an import-export linkage system. Tariff exemptions on raw materials, intermediate goods and capital equipment to be used in exporting goods as well as a tariff drawback system were created. A very generous *wastage allowance* allowed exporters to earn profits on the domestic market, thus encouraging participation in the export drive. The most important assistance granted to exporters was special export financing granted by government-controlled banks to Korean companies which could provide export letters of credit as collateral.

The actual performance of the nation's economy during the first EDP exceeded expectations. GNP rose more than planned (8.5% actual versus 7.1% planned). Commodity exports rose to US$249.5 million in 1966, a 700% increase from 1960. The nation's GNP increased from W246,690 million in 1960 (US$87 per capita) to W382,500 million (US$122.5 per capita) in 1966. Korea's population growth rate dropped from 3.0% in 1960 to 2.75% in 1966. The rapid growth of labour-intensive light industry created a large number of new jobs, reducing unemployment from over 20% in 1960 to 6.8% in 1966. Primary industry grew by 5.3% a year, secondary industry by 15% and tertiary industry by

8.1%. Electric power generation capacity doubled and coal production significantly increased.

The Second Five-Year Economic Development Plan 1967–71

The second EDP was much more fully developed than its predecessor as Korean economic planners became more sophisticated and had more time for preparation. Its basic objective was to "promote the modernization of the industrial structure and to build the foundations for a self-supporting economy." The second EDP was more conservative as a whole than the first EDP, but continued to place a major emphasis on achieving self-sufficiency in the production of foodstuffs. Other goals included the construction of heavy and chemical industry infrastructure and increasing exports. The second EDP also set goals of reducing unemployment and population growth as well as increasing productivity and national income.

Actual performance of the economy again exceeded the targets in almost all major sectors. Annual growth in GNP (9.7%) was greater than the projected levels (7%) and total investment nearly doubled. Export growth exceeded 300% and import growth rose by more than 200%. Total employment also rose substantially during the period. Commodity exports in 1971 were US$1,132 million which represented 16% of GNP. Major export items during the period included plywood, woven cotton fabrics, iron and steel, clothing, footwear and electrical machinery. Manufactured goods accounted for over 85% of commodity exports by the end of the five-year period. The growth of imports during the second EDP prevented any improvement in the balance of payments. Foreign exchange holdings were increased as a result of rising levels of foreign borrowing.

The Third Five-Year Economic Development Plan 1972–76

While the second EDP was intended to improve the economy's structure, the third EDP was directed toward achieving strategic objectives. The main goals of this plan were to continue the growth of industry while at the same time increasing rural development and widening the distribution of wealth and income. The heavy and chemical sectors of industry received particular support during the third EDP in order to compensate for Korea's declining comparative labour cost advantage. Attempts were made to increase regional development and rapid growth was de-emphasized. Instead, efforts were made to increase domestic investment and savings. A number of targets established in this plan were met or exceeded, despite the OPEC-induced oil crisis which occurred during this period. The actual annual average growth rate for GNP was 10.1%, well in excess of the projected 8.6% rate. The level of exports also exceeded expectations rising to US$7.815 billion at the end of the five-year period. The huge rise in imports (a rise to US$8.405 billion) can be largely traced to higher oil prices and the nation's increased energy needs.

The Fourth Five-Year Economic Development Plan 1977–81

The fourth EDP was similar to its predecessors except that it incorporated more private sector input. Major goals for this plan included increasing domestic savings, improving the balance of payments by promoting exports, making improvements related to technology assimilation and creation and increasing the amount of skilled labour. The fourth EDP also began to focus on increasing the nation's level of investment in research and development.

The fourth EDP was the first of the economic plans which did not substantially reach or exceed its goals. Average GNP growth was only 5.5% during the five-year period. Principal factors in causing the disappointing results were the second oil crisis and a disastrous harvest in 1980. Korea had been able to surmount the first oil crisis which occurred during the previous plan only by expanding the economy through increasing the money supply, increasing exports and borrowing more from abroad. By the time the second oil shock came, the economy was already overheated and the inflation level was very high. Imports rose by US$4 billion from 1978 to 1980 because of the increased price of oil and the current account deficit rose to a peak of 23% of total exports during the same period (up from 6.3%).

Civil disturbances caused by the economic distress in some measure contributed to the assassination of President Park. In January 1980, only three months after President Park's death, a number of strong new economic policy measures were introduced to turn the economy around. Interest rates were pushed up sharply, the won was devalued by 16.6% on a trade-weighted basis and domestic energy prices were raised. In March of that year, a two-year standby arrangement was made with the International Monetary Fund for approximately US$800 million in financial assistance. During the first half of 1980, financial restraint was maintained, by continuing to slow the growth of private sector credit, by deferring investment projects and by reducing government expenditure. The won depreciated about 30% on a trade-weighted basis and an estimated 10% after adjusting for relative inflation rates. Korean heavy industry was restructured to solve the problem of earlier overinvestment followed by the sudden cessation of investment. Actions were also taken to improve the equity and efficiency of the tax system and to reduce the nation's dependence on imported oil.

In 1980, Korea's GNP fell 6.2%, the unemployment rate rose to 6.3% and real wages and consumption fell. Fixed investment fell sharply, down 14.8% in 1980. Wholesale prices rose by 38% compared with 19% in 1979. Three-quarters of the price increase was attributable to the effects of energy price increases, the won depreciation and crop failures. The current account deficit continued to increase, totaling US$5.4 billion in 1980 or 23% of receipts. However, export growth began to recover, led by manufactured goods, especially iron and steel and electronics, while the volume of imports fell by 4.3 % due to the recession. Exports in the last year of the fourth EDP were US$20.8 billion. By 1981, the Korean economy had begun to recover and the stage was set for continued improvement.

The Fifth Five-Year Economic Development Plan 1982–86

The fifth EDP's primary goal was the "sustained growth of national power." It placed a greater emphasis on stability than government planning. Some of this change can be traced to failures in the fourth EDP. The fifth EDP called for improvements in the standard of living, defence, and Korea's place in the world community. Because the Korean economy was increasing in complexity, the role of government planning was to be reduced during the plan period. A major focus of the fifth EDP was the reduction of the level of inflation. The phrase "Social Development" was added to the plan to reflect the fact that increases were targeted for housing, health care and other social welfare sectors. The average growth of 7.6% in GNP was to be led by fixed capital investment growing at 9% per annum and exports growing at 10.9% per annum. Consumption, by contrast, was expected to grow more slowly, at 5.3% per annum over the whole period. Meanwhile inflation, measured by the GNP deflator, was targeted to decelerate to single figures by 1986, averaging 10.8% for the period.

The performance of the economy under the fifth EDP was excellent. Inflation was reduced to 2.3% at the end of 1988 and growth targets for GNP were exceeded since per capita GNP had risen to US$2,296 at the end of the plan. The most auspicious development for the economy was the current account surplus of US$4.617 billion. Exports in 1986 were US$33.913 billion and imports US$29.707 billion. The economy began to reach a more mature and stable level during the period and finally began to establish a base for the long term.

The Sixth Five-Year Economic Development Plan 1987–1991

The emphasis of this plan is on "improving the stability and efficiency of the economy". "Growth through stability" is its principal slogan. This will be achieved in part by increasing competition and by gradually decreasing the government's involvement in the economy. The government will also take steps to reduce uncompetitive and unfair practices in the domestic economy. The economy will be gradually internationalized so as to subject domestic industries to increasing competition.

Actions will be taken during the sixth EDP to increase government efficiency through bureaucratic consolidation and privatization of government owned businesses. The plan also includes calls for greater investment in social development (such as housing and education) and balanced geographical development and wealth distribution. Increased levels of resources will be devoted to improving technological capacity and worker education. The level of investment in science and technology should rise to 25% in 1991. The domestic savings ratio is due to rise from 28.4% in 1985 to 33% in 1991.

The sixth EDP is already exceeding its targets as the rate of growth in 1987 was 12.2%, well above the projected rate of 7.7%. Per capita GNP rose to US$2,850 versus the expected US$2,474 and exports were also above projections during the its first year. The only real trouble spot for the government is inflation which was 5.8% in 1987, nearly double the projected 3% level.

Chapter VIII
Financial institutions

The development of financial markets in Korea has been both fostered and retarded by government controls. While the government's policy objectives have largely been achieved, financial markets in Korea have a considerable distance to travel before meeting the standards which exist in major financial centres. As financial sector decontrol and liberalization accelerates during the late 1980s and early 1990s, money and capital markets should expand rapidly in size and sophistication, in a step-by-step programme though, rather than a *big bang*. If the projections of the KDI are correct and Korea has a cumulative current account surplus of US$126 billion in 2010, Korea will be playing a significant role in world financial markets long before then. Some of the major components of Korea's financial market are discussed below.

Commercial banks

Commercial banking in Korea began in 1878 when the First National Bank of Japan opened an office in the port city of Pusan. Other Japanese and Korean banks soon followed. In 1909, the Bank of Korea was established to act as the nation's central bank and to issue currency. The Bank of Korea was renamed the Bank of Chosun in 1911. After Japan's colonial occupation began in 1910, a number of banks were established to finance projects associated with Japan's colonial activities as well as to purchase Japanese government bonds.

In an effort to strengthen the nation's financial system after the Second World War, the National Assembly drafted legislation creating the Bank of Korea. Unfortunately, the Korean War broke out approximately two weeks after the nation's central bank was established and it was not until well after the ceasefire in 1953 that any substantial progress could be made to improve Korea's financial structure. The major early efforts involved forming the Korea Development Bank and the Korea Agricultural Bank as well as strengthening and reorganizing the commercial banks.

Commercial banks in Korea may be divided into four categories: *city banks, local banks, foreign banks* and *special banks*. All commercial bank activities in Korea are closely regulated and monitored by the Korean government. The demand for funds from these commercial banks exceeded supply until 1986 so that each bank could be selective when choosing its customers. In 1986, Korea's current account surplus created considerable liquidity but the government continued to ration credit and to control interest rates. Commercial banks prefer to lend to large well-known companies and are often reluctant to loan funds to small companies which are less well-known and poorly capitalized. This preference largely arises from the paucity and unreliability of financial information on Korean companies, the government's practice of *taking out* foreign banks from bad loans, and the belief that the Korean government will not let a major Korean company default on a loan.

Interest rates on loans given by commercial banks in Korea do not reflect market demand since the government controls rates and fees (most loans must carry a rate ranging from 10–11% as of May 1988), therefore limiting the profits of these banks. Commercial banks raise funds by borrowing on the international and domestic markets, borrowing from the Bank of Korea and from deposits from the public.

Capital has always been scarce in Korea. Companies are nearly always undercapitalized by foreign standards and must rely heavily on short-term borrowings from commercial banks. This borrowing is typically made in the form of a loan which is rolled over. The customer usually issues a short-term promissory note in return for the loan.

City banks

There are five *city banks* in Korea (Bank of Seoul, Korea First Bank, Commercial Bank of Korea, Hanil Bank and Cho Heung Bank) which are the principal lenders to major companies. There are also two other city banks which are joint ventures between Korean and American companies (Kor-Am Bank) and Korean and Japanese companies (Shinhan Bank which is controlled by Koreans resident in Japan). City banks are so called because of the location of their head office, although they also maintain large domestic branch networks and many overseas offices. The city banks primarily lend funds to large Korean businesses. Because the government establishes low interest rates, city banks cannot attract all of their capital from depositors and are therefore actively involved in the money markets.

While these banks were *privatized* (sold by the Korean government to private investors) some years ago, senior management of each bank is appointed by and responsible to the government. As financial decontrol continues under the Roh government, these banks will increasingly adopt their own internal policies concerning personnel and credit. The city banks generally operate primarily as instruments of government monetary, economic and social policy and in practice cannot genuinely be said to be pursuing private goals or profit. Thus, in many respects the banks are more like development banks than true commercial banks. The domestic commercial banks with partial foreign ownership, KorAm and Shinhan, pursue normal business goals to a significantly greater degree.

When the city banks were privatized, rules were established by the government to prevent large businesses (and particularly chaebol) from acquiring a controlling interest in a bank. Theoretically, no shareholder is allowed to own more than 8% of the outstanding shares of a city bank. In practice, cross-shareholdings and nominee shareholders could be easily used by chaebol and other investors to assume control over a city bank. Given the dominance of the chaebol, continued monitoring and control may be necessary to prevent a group or individual from obtaining control over a city bank.

Korea's seven city banks earned W230 billion in pre-tax profits during 1987. Despite these profits, the city banks are in poor financial condition, operating with thin capitalization and high levels of bad loans. The return on assets of the five largest city banks was only 0.16% in 1987. The government's policy of requiring the banks to issue loans at less than market rates to certain industrial sectors and companies is a major cause. A significant portion of these borrowed funds have been diverted to speculative investments. The total outstanding loans of the city banks increased from W20,370 billion in 1986 to W21,417 billion in 1987. Total deposits increased by 28.9% in 1987 to W23,611 billion in 1987.

The official books of the city banks show a low level of net income, low return on assets, and moderate loan loss provisions. However, it has been estimated that almost W7 trillion of the W21 trillion in these bank's loan portfolios should be considered of doubtful quality if international standards are applied. Other experts believe that approximately 25% of risk assets of the five largest city banks are in default. At present, they are capitalized at W1.5 trillion, but will survive despite their poor financial condition, because they are fully supported by the Korean government. The Bank of Korea has consistently confirmed its intention to assist the city banks in meeting their obligations, enabling them to borrow funds from domestic depositors and international lenders.

The financial liberalization which is now in progress has actually made the predicament of the Korean city banks worse. To date, decontrol has primarily affected non-bank financial institutions and secondary markets. This has enabled borrowers to utilize funds beyond their traditional lending sources. This has had two results. First, decreases in margins and profits on lending by commercial banks has reduced profitability. Second, borrowers with already strained finances are able to go further into debt thus worsening the condition of the lending portfolio of the city banks.

Each Korean company, joint venture or foreign-owned subsidiary which meets specified criteria concerning its size, has a bank which acts as its *prime bank*. While additional secondary banks may be designated, the prime bank must approve loans made by secondary banks. The prime bank system is used by the Korean government as a conduit to impose government economic policy on companies. The underlying premise of this system is that Korean companies lack the necessary sophistication required to manage their own financial affairs. In theory, each Korean company's prime bank should oversee these financial affairs. The prime bank is charged with responsibility for: (i) establishing a limit on financings for operational funds and guarantees; (ii) establishing financial controls; and (iii) establishing sound financial management practices. It has been suggested that President Roh Tae Woo's government will eliminate or loosen the prime bank rules as part of a broad financial deregulation package.

To partially compensate for the low lending margins, fees for financial services are set by the Korean government at rates which are far in excess of international standards. Thus, healthy Korean companies are essentially subsidizing the poor loan portfolios of the commercial banks. This practice will be gradually phased out as financial liberalization continues.

Korean city banks are now endeavouring to adapt to their new role as commercial rather than development banks. As the government increases the autonomy given to financial institutions, city banks will need to become more sophisticated. This challenge will be increased by the financial liberalization which is taking place in other financial sectors. City banks will increasingly move into new areas such as selling commercial bills, selling bonds subject to repurchase agreements, issuing credit cards, factoring, providing trust services and issuing negotiable certificates of deposit.

Local banks

There are 10 privately owned *local banks* which may only open branches in the province or special city in which their head office is located (except for one office in Seoul). Local banks are relatively new to Korea's financial markets since the first of these banks was established in 1967. In general, local banks are more likely than the city banks to make loans to medium and small local business enterprises.

These local banks are: the Chung Buk Bank, the Kyungki Bank, the Kangwon Bank, the Daegu Bank, The Bank of Pusan, the Kyeongnam Bank, the Chungchong Bank, the Bank of Cheju, the Kwangju Bank and the Jeonbuk Bank.

The local banks have far less paid-in capital than the city banks. A relatively high proportion of the assets of local banks is obtained through time deposits. The local banks account for approximately 15% of Korea's outstanding deposits and loans. Deposits in these banks rose 29.1% to

W5,859 billion in 1987 and outstanding loans totalled W3,997 billion. Local banks have fewer investment alternatives than city banks and are therefore much more significant purchasers of debentures as a percentage of assets. For this reason, local banks are important suppliers of funds to the city banks. The interest rates and loan fees charged by local banks are regulated by the Bank of Korea.

Because of the need for increased regional development to more evenly distribute the country's increasing wealth, the role of local banks will be increased substantially in the years to come. This change was one of President Roh's major campaign pledges and his administration has formed an intergovernmental task force to achieve this objective. Local banks will be encouraged to move beyond simply providing funds and services for small and medium industries. Improving the liquidity of local banks was kicked off in 1987 when the local banks (with government support) increased their total collective paid-in capital from W209 billion to W339 billion. Moves will also be made to strengthen the local banks' foreign currency and financial services departments.

Foreign bank branches

The number and profitability of foreign banks has grown steadily since Chase Manhattan opened the first foreign branch in 1967. There are now 55 foreign bank branches and representative offices in Korea. American and Japanese banks have the greatest market presence largely due to the level of trade between Korea and these countries.

Foreign banks in Korea (excluding 10 Japanese banks which do not settle accounts on a calendar basis) earned W93.1 billion in after-tax profits in 1987. This puts the return on total assets at 1.76%. After-tax profits for these foreign banks in recent years are as follows:

1986	W71.5 billion
1985	W87.6 billion
1984	W68.7 billion
1983	W55.4 billion
1982	W49.5 billion

Much of the revenue growth for foreign banks in 1987 can be traced to increased levels of foreign exchange transactions and securities related income. The main factor limiting profitability was the lower level of foreign currency loans.

Branches of foreign banks engage in a range of commercial banking activities, including receiving deposits, extending loans, dealing in foreign exchange and investing in securities. Representative offices of foreign banks are involved in more limited liaison activities. Foreign banks are prohibited by the Korean government from engaging in a number of important activities and therefore concentrate on providing corporate financing rather than retail banking. This focus arises because foreign banks have traditionally had very limited access to local currency funding and a very limited ability to establish multiple branches. Recently, Citibank began to move aggressively into the retail banking area, but other banks are unlikely to follow this lead.

From 1984, the Korean government began to take action to increase the access of foreign banks to local currency funding. These actions have included the right to offer certain certificates of deposit, phased access to the rediscount facilities of the Bank of Korea and some access to the trust business. However, percentage and capital asset restrictions imposed by the Korean government have limited the effect of liberalization. To illustrate this point, the share of won deposits held by foreign banks increased only to 1.7% in 1986 from 1.5% in 1983. Furthermore, foreign bank assets declined from 6.7% to 6.4% of total Korean banking assets in 1987. Thus, most of the capital of a foreign bank branch in Korea is borrowed from the bank's head office overseas. The ratio of won loans to won deposits for branches of foreign banks in Korea was 4.99 in 1986 which is far above world standards. In 1986, foreign banks accounted for 9.2% of the total capital in the Korean commercial banking system. As of the end of 1987, approximately 75% of loans by foreign banks were made in Korean currency.

In making loans in Korea, foreign bank branches may lend the funds using two different methods. The first method is a direct loan of foreign currency pursuant to the Foreign Exchange Control Law and Regulations. The amount of these loans which foreign banks may offer is strictly regulated by the Korean government. The interest rate and fees charged as well as the use of the funds must be approved in each case.

The second method is a loan of won currency which typically requires that the foreign bank branch sell foreign currency to the Bank of Korea on the condition that the identical amount be resold to the foreign bank branch at the expiration of a predetermined period. The currency exchange transaction uses a forward exchange rate which provides the foreign bank with a guaranteed profit margin equal to the cost of funds internationally and the won exchange rate. This type of won currency generating transaction is known locally as a *swap*. Every foreign bank has been allocated an amount which may be swapped at any given time. The lending rate for funds generated through swap transactions is set by the Bank of Korea based upon the discount rates on commercial bills utilized by domestic banks. A swap can also be described as a loan of foreign currency by the foreign bank to the Bank of Korea which loans won to the foreign bank, which in turn lends to its customers.

Since Korea's interest rates are substantially higher than those which prevail on international markets, the Korean government has closely regulated the margin which foreign bank branches may earn on swap-generated funds. This allows the Bank of Korea to regulate the profitability of foreign bank branches. The amount of funds available to foreign banks will be reduced as they are allowed increased access to local funding sources. Whether the two changes will balance each other remains to be seen. Many foreign bankers are sceptical regarding the government's intentions. In November 1987, the Ministry of Finance announced reductions in the amount of funds available to foreign bank branches and that the swap system would be entirely phased out. In 1987, the cumulative amount of swaps was reduced by 10% to US$1.69 billion. Compensation for this loss of funds through access to rediscount facilities for financing commercial bills and issuing negotiable certificates of deposit has not been any real assistance. In practice, a foreign bank must increase its lendings to small and medium sized industries from 25% to 35% in order to rediscount commercial bills, and interest rate limits make it difficult to issue negotiable certificates of deposit. As of the end of 1987 no foreign bank had discounted a commercial bill. The Foreign Bankers Group has stated that any reduction in swap lines held by foreign banks must be accompanied by the introduction of a real money market in which foreign banks can freely participate. This means, among other things, market interest rates and the ability to issue certificates of deposit and commercial paper.

Among the difficulties faced by foreign bank branches are the restrictions which the Korean government has placed on their ability to obtain real property security. In essence, the Alien Land Acquisition Law prohibits a foreign individual or company from owning real estate unless it is used for certain permitted business purposes. Because of this restriction, foreign banks have not been able to take a security interest in real property when making loans. As a result, foreign bank branches have not been able to engage in a substantial amount of asset-based financing. This explains the widespread use of bank-guaranteed financing techniques in Korea.

Lending won currency to local companies is unlikely to be as good a source of profits in the future, so foreign banks are looking for profit opportunities in capital markets and other niches. As an example, foreign banks began offering an innovative foreign exchange option to major exporters which created US$650 million in new credits within 17 days. Foreign banks will look to foreign exchange trading and letters of credit as well as securities issues to increase profits. As much as 50% of the profits of some foreign banks is attributed to foreign exchange dealings.

The long-term policy of the Korean government appears to be directed toward restricting the role of foreign banks in Korea without harming the nation's ability to borrow again if the current favourable economic conditions change. Reducing the number of foreign banks is likely to be accomplished by attrition rather than revoking licences and only a few applications to establish foreign bank branches can be expected to be approved each year. New branches are no longer given access to swap funds.

As Korea's foreign debt is reduced by the continuing trade and balance of payments surplus, Korea will have less need for foreign banks as a source of foreign capital. To give domestic banks room to recover their financial health, the Korean government is doing everything possible to restrict the share of foreign banks in the domestic market. The Korean government will be walking a tightrope between the pressure exerted by foreign businesses and governments and the cries of domestic political groups who want to limit profits by foreign banks. In general, the Korean government has been "liberalizing" by imposing the same restrictions on foreign banks which made domestic banks financially unsound. Foreign banks particularly want to avoid participation in the financial restructuring of failing Korean companies as this will considerably reduce their profitability. Because of this trend, several foreign banks have decided to downgrade their operation from a branch to a representative office or close their Korean presence entirely.

The central and special banks

The Bank of Korea has been the country's central bank since 1950, responsible for establishing both credit and monetary policy. The Bank of Korea is governed by the Monetary Board, whose members are appointed by the government and which is chaired by the Minister of Finance. The current Monetary Board consists of six economics professors, a retired banker and the governor of the Bank of Korea. The bank is responsible for issuing currency as well as supervising the activities of all other banks in Korea. The new governor, Kim Kun, promised in early 1988 that commercial banks would be granted more autonomy.

Revisions to the structure of the Bank of Korea are now being debated. Many would like to see the bank with a significantly greater degree of independence from the government. These advocates believe that this change would insulate Bank of Korea policy from purely political manoeuvering and increase the stability of the Korean economy. Others, including the Minister of Finance, Sakong II, argue that monetary and fiscal policymaking functions should not be separated. The Governor of the Bank of Korea and many top employees of the central bank are pushing for a more independent central bank. This debate has existed in some form since 1962, when Governor Yoo Chang-Soon resigned in protest over moves to reduce the Bank of Korea's independence. Most recently, the March 1988 announcement that Governor Park Sung Sang had been dismissed and replaced by Kim Kun was an indication of the government's desire to maintain firm control, despite serious criticism. Park Sung Sang's testimony before the National Assembly in 1987 requesting increased autonomy is widely regarded as the reason for his replacement.

Since the won may appreciate by as much as 20% against the US dollar and because of the interest rate differential, the foreign exchange holdings of the Bank of Korea are losing approximately 2% of their value each month. Repaying the foreign debt is helping, but at current rates Korea will be a net creditor in the early 1990s. Government economic planners wince at the cost to the bank of simultaneously offering below market rates to commercial banks and reducing liquidity by issuing Monetary Stabilization Bonds.

Another drain on the Bank of Korea are government efforts to resuscitate failing companies by obliging city banks to participate in restructuring and reorganizations. This markedly increases their level of non-performing assets and the Bank of Korea must then support them by making concessionary loans. In June 1987 the bank lent over W730 billion at an interest rate of 3% (which was less than the prevailing rate of inflation).

The Bank of Korea supervises the following six non-depository commercial banks which were created in the 1960s to achieve certain policy objectives.

The Korea Exchange Bank (KEB), which specializes in providing international banking services, including foreign exchange and trade financing. The KEB has established a network of over 30 branches in foreign countries. This bank is no longer considered by many to be a specialized bank, although it does provide some services not available elsewhere.

The Small and Medium Industry Bank, which provides funds to these companies, principally in the manufacturing industries.

The Citizens National Bank, which provides short-term credit to individuals and small companies. This bank was originally formed because the services of commercial banks were not available to small customers.

The National Agricultural Cooperatives Federation, which through its credit department makes loans related to agriculture and forestry.

The Central Federation of Fisheries Cooperatives, which through its credit department makes loans to the fishing and aquaculture industries.

The Korean Housing Bank, which supplies housing loans to low and middle class families. This bank was established in 1967 by the government. In 1987, the Korean Housing Bank held an 88% share of the institutional private housing finance market in Korea. The Korea Housing Bank has a nationwide network of over 180 branches.

Several other special banking institutions exist which are government funded and non-depository. These institutions make loans on projects which are deemed important to the economy. These banking institutions include:

The Korean Development Bank, which provides long-term credit for major projects in the areas of public works, transportation, communication and industry. This wholly government owned bank has been the major institution through which Korea has obtained long-term capital. Much of this bank's funds come from major international lending institutions such as the World Bank and the Asian Development Bank and from floating bonds on international markets.

The Export-Import Bank of Korea, which finances all aspects of foreign trade and overseas investment, as well as extending credit to foreign governments, banks and industrial corporations.

The Korea Long-Term Credit Bank (formerly the Korean Development Finance Corporation), which promotes private enterprise through medium- and long-term industrial loans (often on the basis of equity participation) usually in foreign currencies. This bank was established in 1980 as a joint venture between Korean and foreign investors. It is modelled on the very successful Japan Long-Term Credit Bank. Most transactions involve equipment and facilities investment. Funds raised by the Korea Long-Term Credit Bank are loaned to private companies.

Korea also has a number of savings and non-banking financial institutions. Investment and trust companies serve an important function, growing by 45% during 1987. Thirty-two short-term finance companies have been created by the private sector with government encouragement to serve the same function as the commercial sector of the *kerb* market.

The outlook for the short-term finance industry is not good, however, as government-imposed restraints to reduce liquidity have made it difficult for short-term finance firms to raise capital. Fortunately, most of them have been able to make profits in the capital markets. As of December 1987, short-term finance companies had total deposits of W9,675 billion and grew at 13.5%. The largest companies are Daewoo, Daihan, Hanyang, Central Seoul, Chaeil and Goldstar, each with paid-in capital of W30 billion. Many in government believe that short-term companies have outlived their usefulness and will soon be merged into banks and securities companies issuing debentures.

Mutual savings companies have been created to serve the consumer side of the kerb market. The Korean government has also allowed the formation of credit unions and mutual credit associations as well as establishing a postal savings system.

Merchant banks

The Merchant Bank Act was enacted in order to foster the diversification of Korea's access to foreign and domestic capital and to increase the supply of medium- and long-term funding. The first merchant bank was established in 1976 and by 1987 six merchant banks were operating in Korea. Each of these banks is a joint venture with one or more foreign financial institutions or investors. These banks are: Korea Merchant Banking Corporation (a 50:50 venture between Korean and UK investors), Korea Kuwait Banking Corporation (a joint venture between Korean and Kuwaiti companies), Saehan Merchant Banking Corporation (a joint

Financial institutions in Korea

- **Central Bank**
  - The Bank of Korea

- **Banking Institutions (Deposit Money Banks)**
  - Specialized Banks
    - Korea Exchange Bank
    - Small and Medium Industry Bank
    - Citizens National Bank
    - Korea Housing Bank
    - National Agricultural Cooperative Federation
    - National Federation of Fisheries Cooperative and Member Cooperatives
    - National Livestock Cooperatives Federation
  - Commercial Banks
    - Nationwide Commercial Banks
    - Local Banks
    - Foreign Bank Branches

- **Non-Bank Financial Institutions**
  - Saving Institutions
    - Trust Accounts of Banking Institutions
    - Mutual Savings and Finance Companies
    - Credit Unions
    - Mutual Credit
    - Postal Savings of Post Offices
  - Development Institutions
    - Korea Development Bank
    - Export-Import Bank of Korea
    - Korea Long Term Credit Bank
  - Investment Companies
    - Investment and Finance Companies
    - Merchant Banking Corporations

- **Securities Markets**
  - Insurance Companies
    - Life Insurance Companies
    - Postal Life Insurance of Post Office
    - Securities Supervisory Board
  - Korea Stock Exchange
  - Securities Companies
  - Korea Securities Finance Corporation
  - Securities Investment Trust Companies

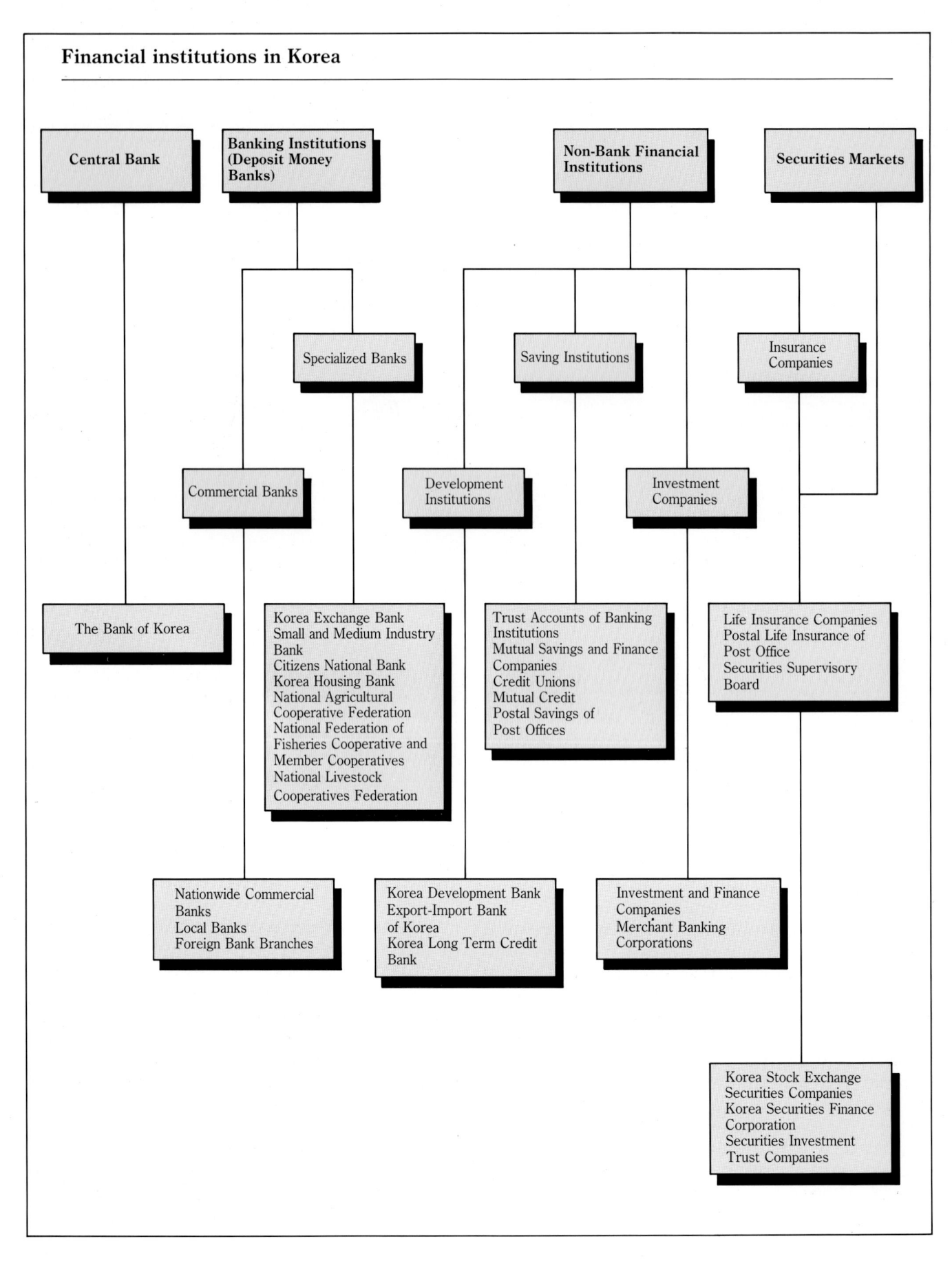

venture involving Korean, US, Japanese and UK investors), Korean French Merchant Banking Corporation (a 50:50 venture between investors from those two countries), Asian Banking Corporation (a 50:50 venture between Koreans and a Saudi Arabian company), and Korea International Merchant Bank (another 50:50 joint venture with Hong Kong and German investors holding the foreign shares). Merchant banks in Korea offer the following services:

- Short-term won financing
- Short- and medium-term foreign currency loans
- Foreign exchange services
- Corporate advisory services
- Bonds and guarantees
- Leasing
- Corporate debenture issues
- Certain securities-related businesses
- Short-term investment trusts and cash management accounts.

To date, Korean merchant banks have concentrated on providing short-term financing and have primarily raised funds through the issuance and discounting of short-term promissory notes. As much as 70% of the operations of merchant banks involves short-term commercial paper, but as these banks gain more experience, more medium-term financing will occur. Merchant banks have in the past had to pay a premium in the interbank market in raising funds. Despite this increased cost, merchant banks have had a better return on earnings than foreign bank branches and domestic banks. Merchant banks have experienced substantial growth since their establishment. The government announced in late 1987 that Korean merchant banks may open overseas offices.

Other financial institutions

Leasing firms

Eight specialized leasing firms have been established in Korea. These firms are: Korea Industrial Leasing, Incorporated; Korea Development Leasing Company, Limited; First Citicorp Leasing Incorporated; Hanil Leasing Company, Limited; the CNB Leasing Company, Limited; Pusan Leasing Company, Limited; Korean Leasing Corporation; and Taegu Leasing Corporation. Four of these companies are joint ventures with foreign institutions (the first four). The Korean leasing industry is relatively young, since leasing companies came into existence only in the early 1970s. Korean firms concentrate on finance leasing as the markets for operating leases, sub-leases and leveraged leases have yet to be developed in Korea. A majority of the leases involve industrial machinery, office automation equipment and transportation equipment. Leases are generally structured on a floating rate basis and denominated in a foreign currency (exchange rate risk is borne by the lessee). The volume of lease contracts rose from W122.1 billion in 1980 to W1,423.6 billion in 1986. The KDI has predicted an annual growth of 20-30% for the leasing industry over the next several years.

The chronic shortage of funds is the primary cause for the growth in the Korean leasing industry. Ironically, it is the same shortage of funds which constrains the growth of the leasing industry. This means that leasing companies will grow directly in proportion to the amount of capital allocated to it by the government. Like companies in foreign countries, Korean companies have primarily been attracted to leasing because it does not add debt to their balance sheet. Leasing is also attractive to smaller firms which cannot offer collateral to lenders. For equipment financing, a lease transaction requires no down payment (in contrast only 80% of the purchase price is typically available in the case of a loan). Lease financing in Korea also typically does not require collateral security.

Merchant banks compete with the specialized leasing companies for business. In general, merchant banks concentrate on large transactions involving very significant pieces of equipment while leasing firms lease all types of equipment. In providing leasing services, merchant banks have the advantage of being able to deal in foreign exchange and provide other related services. However, the specialized leasing firms were responsible for over 80% of all leases in 1986.

As is the case with all Korean financial institutions, the demand for funds exceeds the available supply. Leasing firms raise funds by borrowing from abroad and from domestic banks and by issuing debentures. While authorized to do so, Korean leasing firms have had difficulty in issuing debentures. As the industry becomes more financially sound, this difficulty will disappear.

Insurance firms

The insurance industry in Korea is divided into life and non-life sectors. No insurance company doing business in Korea may engage in business in both sectors. Although the insurance industry has existed since the Japanese colonial period, significant growth did not occur until the 1970s. There are six primary Korean life insurance companies. In December 1987 the government announced a plan to allow the formation of life insurance companies in regional areas.

Korean life insurance companies are relatively unsophisticated and are only now beginning to market attractive policies. To illustrate problems in the industry, around 75% of all policies are allowed to lapse as the initial term expires. The Korean government sets premium and interest rates for all life insurance policies. Approximately 38% of the Korean population is covered by a life insurance policy. As of the end of March 1987 (the end of fiscal year 1986), the

Korean life insurance industry had over W100 trillion of policies in force. Statistics released by the Life Insurance Association of Korea show average annual growth of around 30%. The combined assets of Korean insurance companies passed the W9 trillion mark as of the end of the 1986 fiscal year. The total amount of aggregate premiums makes the Korean life insurance market the seventh largest in the world.

Despite impressive growth in recent years, profits in the life insurance industry have been low. The industry as a whole registered its first collective profit during the 1986 fiscal year (W1,574 million). Poor financial management has been a major contributor to low profitability in the life insurance sector. A major contributor to the lack of profitability are government regulations concerning asset management. The government requires that life insurance companies invest excess funds in a conservative manner. Furthermore, life insurance companies must purchase certain amounts of government bonds. To make matters worse, a significant proportion of excess funds must be loaned to small and medium industries. These life insurance companies contribute to their own financial problems since a good portion of their available capital is loaned to affiliated companies (chaebol).

Foreign life insurance companies have recently been allowed limited access to the insurance market in Korea. Life Insurance Company of North America (LINA), American Life Insurance Company (ALICO) and American Family Assurance Company (AFLAC) have established branches in Korea. In an executive agreement entered into between Korea and the US, the Korean government has agreed to allow even greater access by US companies to Korea's life insurance market. The Korean government has stated that it will allow joint ventures in the insurance industry as long as they do not involve one of the fifteen largest business groups. Aetna Life and Casualty Company has announced plans to form a joint venture with Dongbu and Metropolitan Life Insurance Company plans a joint venture with Kolon. It has been reported that a number of Korean securities firms wish to enter the life insurance industry through joint ventures. The interested Korean securities companies include Dougsuh, Tongyang, Hanshin and Hangang. Chaebol ranked between 15 and 30 in terms of consolidated assets are required to hold less than 50% of the equity in any joint venture in the life insurance sector. The government will require that 30% of the paid-in capital of a life insurance joint venture be deposited with the Korean Insurance Corporation as security against a possible default.

The dominant Korean life insurance company is Dongbank Life which holds nearly 40% of all individual and group coverage in Korea. Dongbank Life has nearly W38 trillion worth of insurance in force with over 4.5 million policyholders.

Korea also has a non-life insurance industry. The major casualty and liability companies are Haedong Fire and Marine, First Fire and Marine, Daehan Fire and Marine, Aukuk Fire and Marine, Hyundai Fire and Marine, Shindong A Fire and Marine and Korea Reinsurance. Foreign insurance firms have been allowed to sell certain types of fire, accident and other casualty insurance since 1977. However, foreign companies have not been allowed to participate in the compulsory and *pool* insurance areas. Under Korean law, only 11 Korean companies have been granted a license to sell compulsory fire insurance.

Korea's largest commercial banks			
	Total assets ($ billion)	Net profit ($ million)	Net worth ($ million)
Korea Exchange Bank*	25.2	37.0	922
Commercial Bank of Korea	13.8	9.2	480
Hanil Bank	12.7	18.6	469
Korea First Bank	12.7	16.9	450
Cho Hung Bank	12.6	11.0	429
Bank of Seoul	11.3	13.5	439

*Considered a specialised bank for regulatory purposes,
Note: Figures are for the year ended December 31 1987 converted at 760 = won $1.

Chapter IX

The money markets

The formal money market

Korea has a fairly sophisticated money market which includes short-term deposits, commercial bills, commercial paper and treasury bills. Non-residents are prohibited from investing in the money market. The call deposit market in Korea is limited to interbank transactions with a relatively high turnover. The lending periods vary from overnight to several weeks. Time deposits are accepted by all depository banks at terms which vary from one to twenty-four months. Interest rates on all types of deposits are controlled by the government. The spread between deposit rates and lending rates is very high by world standards. Rates for deposits of over one year are significantly higher than rates available for short-term deposits.

Commercial bills and commercial paper are widely used by Korean companies. Commercial bills are utilized to finance the short-term needs of these companies with terms as long as six months. Banks or other financial institutions have accepted most commercial bills and as a result an active secondary market for these instruments exists. Accepted commercial bills are eligible for rediscounting with the Bank of Korea. Commercial paper is generally issued in the form of promissory notes with a 180-day maturity. Commercial banks generally act as intermediaries in the issuance of commercial paper. In most cases, an underwriting syndicate is established to place the issue. The government does not regulate the rate of interest paid on commercial paper.

Treasury bills are issued approximately four times a year by the Bank of Korea and their maturities range from 30 to 91 days. The government offers treasury bills at a discount in making an issue and uses the proceeds from sales to finance a variety of short-term needs. There is a secondary market in treasury bills.

The informal money market

Because access to capital is so closely controlled by the Korean government and because it sets interest rates at a level which does not reflect the market, a large informal kerb market for capital has developed. The activities of this market are illegal but generally tolerated by the government. Government efforts to regulate and tax income from this market have never been successful. Since it has no official home, early brokers stood on the street at certain locations, giving the market its curious name. The market is served by a network of brokers operating principally in the Myongdong area of Seoul. The funds which comprise this market are generated by groups of small individual investors who form investment trusts (known as *kye*) to increase their otherwise meagre investment opportunities. Traditionally, the kye have been organized and operated by women. Korean men have always entrusted the management of household funds to women since Confucian abhorrence for money and greed made this task unsuitable for a *refined* man. Transactions on the illegal money market can be separated into two types. The first type provides short-term money to businesses for use as working capital. Much of this lending is conducted by brokers on a *call* basis. Brokers keep a tight watch on the finances of all borrowers. Maturities on this type of transaction seldom exceed three months. Brokers have charged borrowers from 5-6% per month and have paid their lenders 3-4% per month, depending upon the risk involved, in any given loan.

The second type provides consumer loans to individual borrowers. This segment of the market is smaller and less well organized than its commercial counterpart. Rates on this section of the kerb market can be double or triple those on official markets. This second type of financing has served as an important source of capital for very small businessmen. Many of these businessmen would have been unable to secure loans despite being creditworthy because it was prohibitively expensive for financial institutions to investigate their credit standing. Lenders in the informal kerb market are likely to already know the borrower and his credit status. This results in lower transactional costs and increased capital availability for small borrowers.

Rates in the kerb market have decreased in recent years due to financial market liberalization. Borrowers on the kerb market provide promissory notes to the lender (with face values which are 2 to 5 times the amount borrowed) as security and pay interest on a monthly basis.

Major financial and political scandals involving the kerb market took place in 1982 and 1983, involving attempts to create a secondary market and major fraud. These scandals were a major embarrassment to the government of President Chun and have to a degree stunted the growth of this market. The Korean Economic Institute estimated the size of the kerb market exceeded W1,100 billion at the end of 1981. Rather than set a dollar figure, some knowledgeable observers have guessed that the kerb market has a 40% share of the total volume of loans outstanding in Korea. A report issued by the Citizen's National Bank in 1985 showed that borrowing from the kerb market was W535,000 per household, only slightly less than the W704,000 borrowed officially. Combining this amount with private corporate borrowings of W1.2 trillion and farm sector private borrowings of W1 trillion, the total kerb market would have been around W5 trillion in 1985.

The size of this illegal market should decrease with financial liberalization. No substantial secondary market for promissory notes issued on the kerb market exists in Korea and no such market is expected to develop because of the major political scandal of 1982.

The stock market

The Korea Stock Exchange

The Korea Stock Exchange was established by the Ministry of Finance in 1956 in Seoul. Prior to that time, the only securities trading occurred within small groups of corporate leaders and major entrepreneurs known as the *Securities Club*. While significant progress has been made since its inception, the Korean stock market is still developing due to several factors. Most importantly, Korean companies are reluctant to sell shares to *outsiders*, since doing so could result in a loss of control over their management and finances. Korean companies are also reluctant to comply with the disclosure requirements associated with a public listing and to pay dividends to shareholders. Even so, during 1987, 31 new companies issued shares to the public and by the end of 1987, 387 companies were listed on the Korea Stock Exchange. Total market capitalization exceeded US$30 billion.

Another factor restricting growth was the reputation of the stock market as a den of speculators. So-called *big hands* (individuals with large amounts of cash) are generally believed to have a huge influence on stock market movements. They and other speculators also engineer large swings in stock values by spreading rumours and insider trading. One of the four major stock crashes which have caused public confidence to falter occurred as a result of the government's efforts to sell shares of the Korean Stock Exchange on the exchange itself. Just prior to the crash, speculation in Korea Stock Exchange share certificates caused prices to rise extremely rapidly, but when the necessary legislation was not enacted after all, the price of nearly all shares on the market crashed.

Another limiting factor is a shortage of experienced brokers and managers and a general lack of public knowledge about the way the stock market operates. Most Koreans continue to view the stock market as a place to reap short-term speculative profits from price swings. This increases instability and thus harms investor confidence. Investor confidence in the stock market is fragile since approximately 90% of individual investors have been in the market for less than one year. If the market turns down substantially, a mass exodus from equities may result, producing a price *meltdown*.

From 1972, the Korean government finally began to take the actions needed to stimulate the equity market in Korea. Legislation was enacted to encourage and even require companies to go public and insider trading was prohibited. As part of the government's effort, the Ministry of Finance began to closely monitor the securities market. The share market recovery was also supported by revisions to the Corporate Income Tax Act which created incentives for companies to go public and penalties for those which do not sell shares. Because of these actions, in 1977 alone the number and value of listed shares on the Korea Stock Exchange doubled. Despite steps taken so far by the government, further action is necessary to strengthen the market. Most importantly, the financial disclosure system needs to be improved. Also necessary are increased efforts to penalize individuals engaging in insider trading and an investor education campaign.

The strengthening of the Korean Stock Exchange was undercut considerably by the OPEC oil shock of 1979; a slower economy and the assassination of President Park Chung Hee in 1980; and the 16.6% devaluation of the won which was intended to counter a major economic slowdown.

Another impediment to the development of the Korean securities market was the government's strict regulations on capital formation and movement. In order to achieve the objectives contained in the five-year EDPs initiated in 1961, the government allocated much of the country's available capital to specific sectors of the economy. To encourage companies to devote their resources to that sector, the capital was offered in the form of long-term loans at below market interest rates. Since most of the targeted areas involved heavy industry and large-scale infrastructure projects, the capital was allocated to large companies with the expertise needed to complete and manage these projects. This did not leave enough capital to form a healthy securities market, and the remaining capital was invested instead on the kerb market for very high returns. The larger the companies became, the less desirable it became for them to issue shares, and the small- and medium-sized had no ability to do so.

As of 1988, fewer than 50% of the largest 15 company groups (chaebol) in Korea have listed their shares on the Korea Stock Exchange. The value of companies listed on the Korean Stock Exchange represent only 30% of GNP at the end of 1987 according to Ministry of Finance statistics. In the US, listed companies produce approximately 50% of GNP. The total capitalization of the Korean securities market is the equivalent of US$37.5 billion.

Korean institutional investors accounted for approximately 70% of share volume in 1987. Korea currently only has approximately 2.3 million citizens who own listed shares.

With more than 410 companies listed on the Korea Stock Exchange and a total market capitalisation of US$42.6 billion, the market is now the 17th largest in the world, and one of the most promising

Korea's police are well trained and secure a stable environment for business

The government is taking steps to raise this figure to 5 million by the end of 1988 and 10 million over the next several years by selling stock in government-owned corporations, establishing employee stock ownership plans and allocating shares to low and medium income individuals. The government plans to sell shares with a value of W1.2 trillion in Pohang Iron and Steel Co. (POSCO), Korea Electric Power Company (KEPCO) and Citizens National Bank as soon as possible. Other government issues may total as much as W332.9 billion in 1988 and W755.5 billion in 1989.

Over the next five years, the government will sell shares in Korea Telecommunications Authority, Korea Exchange Bank, Korea Development Bank, Korea Appraisal Board, Korea General Chemical Corporation, and Korea Monopoly Corporation. These shares will be offered preferentially to low-and middle-income individuals who will be required to open special accounts with securities firms to receive the government-owned shares. This is intended to facilitate contacts between them and the securities firms.

The government wants 500 companies to have listed shares at the end of 1988 and over 1,000 Korean companies to have listed shares by the mid-1990s. To achieve this goal, the government wishes to channel the funds currently used for real estate speculation to the share market. For this reason, the government will continue its vigorous efforts to stamp out real estate speculation. Financial institutions, such as the KorAm Bank and Shinhan Bank, will be encouraged to issue shares to the public during 1988. In early 1988, the Korean government took new steps to increase the number of shares listed on the Korea Stock Exchange. More than 75 companies with aggregate debts exceeding W50 billion have been ordered to retire 5% of their outstanding debt. Similarly, 110 companies with an aggregate debt of W20-50 billion will be required to retire debt by making equity offerings.

One peculiar feature of the Korean stock market is that only approximately 25% of shares are held by individuals under their real names. Individuals open accounts under assumed names to avoid taxes, and to hide their actions from the government and creditors, among other reasons. Some of the money sent to Korea by foreigners uses this false name system to hide the origin of the funds.

During 1987, the Korean Composite Stock Price Index (CSPI) rose to 525.11, which represents a 95% increase from the end of the previous year. The increase for the five-year period ending December 1987 was 330%. It was heartening for investors that the Korean CSPI fell by only 3% in the two weeks after Black Monday. The total market value of shares on the KSE increased by 118% to W26 trillion in 1987. At least fifty new companies are expected to list shares on the Korea Stock Exchange during 1988. Since the P/E ratio on the KSE is low by world standards (10.6 x) and since earnings growth should be up by at least 30% in 1988 (down from approximately 50% in 1987), the prognosis for a rise in the market is excellent. It should be noted that P/E ratios are 30-50% of those in other Asian NICs. The lack of other good investment alternatives should also boost the stock market.

In the past, and for the foreseeable future, the Korean government through the Ministry of Finance will continue to closely regulate the health of the Korea Stock Exchange. By altering the margin requirements for securities companies, changing the credit offered by the Korea Securities Finance Corporation, changing lending policies toward Korean companies by banks, and issuing shares from government-owned corporations, the government will move to keep the market from falling too severely or becoming overheated. The Korean government fully recognizes that investor confidence in the securities market is tenuous. Dramatic swings in stock prices can only undermine this confidence. The government also believes that the middle and lower classes may view an overheated stock market as another instance of the rich getting richer. Until more individual Koreans own shares, the government is unlikely to allow an extremely large run up in share prices.

Approximately 65% of the Korean Stock Exchange is owned by the Korean government with the remaining shares being owned by the 25 Korean securities firms. The KSE is governed by a Board of Directors, the Chairman of which is appointed by the President of Korea. The exchange trades from 9.40am to 11.40am and from 1.20pm to 3.20pm Monday through Friday and from 10.00am to 3.20pm on Saturday. The address of the exchange is:

> Korea Stock Exchange
> 33 Yoido-dong
> Yeongdeungpo-Ku
> Seoul 150
> Tel.: 783-2271

The most widely followed market index is the CSPI which is computed and published on a daily basis by the Korea Stock Exchange.

Securities trading

Securities are traded on the exchange by Korean securities firms in a brokerage capacity and as principals. Transactions on the share market are classified as either *spot* or *regular*. Stocks are traded on a regular basis while securities are traded on a spot basis. In spot transactions, stocks are delivered and payment made on the contract date. Regular transactions are settled on the second business day following the contract date. Margin trading in securities was introduced in 1981 for regular transactions. A margin purchaser is only required to deposit a certain portion of the purchase price with the securities firm. The remaining funds are borrowed from the securities firm. Margin requirements for the purchase of securities are set by the Securities and

Internationalization

A decision by the Korean government to gradually internationalize the securities market was made in January 1981. The government decided to liberalize the securities market in the following four stages:

Stage 1 (1981-4)
Establish international investment trusts domestically and abroad to invest in Korean securities on a limited basis. Exchange Commission (SEC). Margin requirements will be varied by the SEC from time to time to keep the market from becoming depressed or overheated. Shares are traded in units of 10 shares.

Securities transactions in Korea are governed by the SEC pursuant to the Securities and Exchange Law and Regulations. Certain provisions in the Korean Commercial Code supplement the Securities and Exchange Law. Korean securities laws place restrictions on mergers and takeovers, protect minority shareholders in certain situations, penalize individuals and companies which issue false information or fail to make adequate disclosures and penalize certain types of insider trading. A company listing shares must enter into an agreement with the Korea Stock Exchange in which it agrees to comply with certain requirements. Companies listed on the Korean Stock Exchange must also provide annual and semi-annual reports to the SEC.

Establish overseas representative offices of securities firms, and representatives offices and foreign securities houses in Korea.

Stage 2 (1985)
Permit direct foreign investment in Korean securities on a limited basis, i.e., via the floating of convertible bonds by domestic enterprises in overseas capital markets.

Stage 3 (late 1980s and early 1990s)
Allow direct foreign investment in Korean securities without limitations. Allow the floatation and listing of Korean equity in overseas markets. Permit Korean securities houses to handle overseas business. Permit foreign securities firms to conduct business in Korea.

Stage 4 (uncertain, probably mid-late 1990s)
Permit free flows of capital. Allow portfolio investment in foreign securities by Korean investment trust companies (and others). Allow Korean citizens to invest in foreign securities. Allow foreign securities to be listed on the Korean Stock Exchange.

Although foreign firms and investors are eager to see the Korean stock market opened to non-residents, the government-appointed *Financial Sector Development Committee* is unlikely to develop a specific plan. Change will be gradual and made on a step-by-step basis. The most likely action of the government on this issue would be permission for foreign investors to collectively hold a certain percentage (perhaps 15-20%) of the shares of individual companies. Companies in certain industries (e.g., defence, banking) will remain off-limits to foreign investors. This limited opening should occur in 1989 after the situation in the National Assembly stabilizes.

One major problem the government faces in timing the opening is the high liquidity of Korea's financial system. New foreign investment can only weaken the government's moves to control the money supply and inflation. The other major problem is the political situation. Opposition leaders will be quick to wrap themselves in the Korean flag and cry "sell-out" if the DJP liberalizes too quickly. Many of the electorate fear that Japanese investors will control Korean companies through equity purchases. This is precisely why there is no Korean Japan Fund to date. Older Koreans remember Japanese domination of Korean industry and younger Koreans are economic nationalists.

Should any of the opposition leaders become president, there would be little change in economic policies as changing the course of the fastest growing economy in the world would be foolhardy. Kim Dae Jung, and to a lesser extent Kim Young Sam, have limited themselves to making broad statements favouring a more equitable distribution of wealth and a reduction in the economic dominance of the large business groups. The opposition parties could be expected to be more intransigent in foreign trade negotiations to placate economic nationalists.

As an example of how little opposition leaders have considered economic policy, Kim Dae Jung in an interview in May 1988 claimed to be interested in greater independence for the Bank of Korea from the Ministry of Finance. At face value, increased independence for any organization must appeal to Kim Dae Jung. However, the question arises as to whether an independent Bank of Korea is at all consistent with his economic populism. An independent central bank would seriously impede Kim Dae Jung's power to make the changes he desires since its decisions would be based upon what was best for the economy rather than what was politically expedient.

Whether the new US President is a Republican or a Democrat, increased pressure is likely to be applied to the Korean government to open Korea's financial markets. At times in the past, it must have appeared to Korea that US government officials believe that their most strategic future products are cigarettes, beef and chocolate-covered peanuts. Certainly, a new adminstration headed by George Bush or Michael Dukakis would begin to focus on more important industries like financial services and high technology industries.

Four step liberalization plan

Stage	Liberalization Measures

Stage I (1981–1984)	• Permit indirect foreign investment in the Korean securities market.
	  – Five Korean investment trusts marketed outside Korea to non-Korean investors during 1981–1986, totaling Us$140 million.
	  – Shares in two closed-end investment funds, totalling US$130 million, offered to foreign investors during 1984–1987.
	• Permit foreign securities firms to set up representative offices in Korea.
	  – 16 foreign representative offices established in Seoul.

Stage II (Mid 1980s)	• Allow direct foreign investment in the Korean securities market with certain limitations. • Allow the issuance of Korean corporate securities overseas. – 4 convertible eurobonds (Samsung Electronics, Daewoo Heavy, Yukong and Goldstar), totalling US$110 million.
Stage III (Late 1980s)	• Permit direct foreign investment in the Korean securities market without limitations,
Stage IV (Early 1990s)	• Fully liberalize capital movements in and out of Korea.

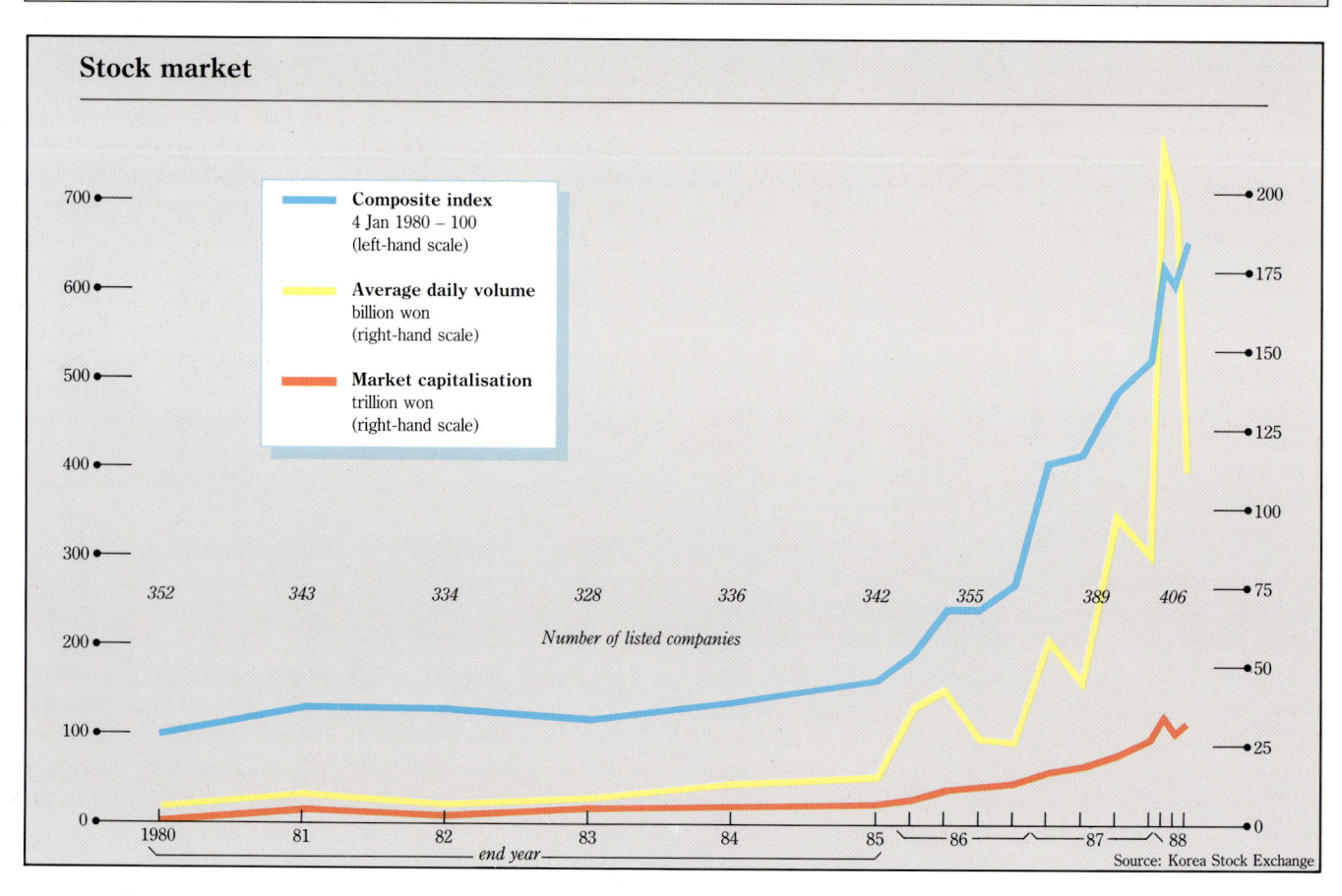

Stock market

Composite index
4 Jan 1980 – 100
(left-hand scale)

Average daily volume
billion won
(right-hand scale)

Market capitalisation
trillion won
(right-hand scale)

Number of listed companies: 352, 343, 334, 328, 336, 342, 355, 389, 406

end year 1980–88

Source: Korea Stock Exchange

Listing

The Korean Stock Exchange is divided into two sections. A listing on the more prestigious first section requires that (i) the company has after-tax profits of at least 5%; (ii) has declared a dividend at lease twice during the past three fiscal periods; (iii) has had their shares listed for more than six months; and (iv) 40% or more of their share capital must be freely traded and held by more than 300 investors. The second section consists of shares which do not meet the above requirements. Even second section stocks must be issued by a company which has been in existence for more than three years and have a paid-in capital of W500 million or more.

Companies which are listed on the Korean Stock Exchange must provide annual and semi-annual financial reports to the SEC as well as report certain major transactions (e.g., mergers, defaults, increase of paid-in capital).

New shares in a Korean company are generally issued at par value to existing shareholders. Shares are also issued at a premium to par value (which may or may not equal market value). The premium is typically set by taking both asset value and projected income into account. The authorities are attempting to encourage a *market price* system of share issuance. To date, Korean investors have not been favorably impressed by market priced issues. Some theorize that major holders prefer the par value system as it requires less capital to maintain a given equity stake. Shares may also be issued by a company to shareholders for no consideration by making a balance sheet addition to paid-in capital. To sell shares to the public, a Korean company must put up at least 30% of its total shares and also reserve 10% of these shares for purchase by its employees.

A few of the Korean companies in which foreign companies are investors have obtained a listing on the Korean Stock Exchange. Korean listed companies with foreign investors include Sun Kyong Fibers, Gold Star Cable, Orion Electric and Han-Dok Remedia. In each of these cases the foreign company owned less than 50% of the equity before the offering. In most cases, the foreign investor(s) retained at least a 33.3% equity share in order to protect certain veto rights and safeguard the value of their investment. Many foreign companies have successfully resisted pressures to go public since this could affect their management control and the government has not been insistent so as to maintain Korea's image as an attractive place to invest funds.

The over-the-counter market

The SEC has issued a regulation which is intended to foster the creation of a formal over-the-counter market in order to allow small and medium sized enterprises to obtain direct equity financing and to provide investors with new investment opportunities. Share trading in the newly organized over-the-counter market began in April 1987. Only shares issued by companies which meet certain specified requirements and which have been registered with the SEC are eligible for sale on the over-the-counter market. Shares issued by small and medium enterprises and high-technology companies which satisfy certain requirements are partially exempt from these requirements. Shares must be registered with the Korea Securities Dealers Association (KSDA) in order to be traded on this market. Both listed and non-listed shares can be traded in the over-the-counter market if they have been properly registered.

Shares traded in the over-the-counter market are generally those which do not qualify for listing with the Korea Stock Exchange and usually involve a significantly higher degree of risk. A Korean company which wishes to trade its shares on the over-the-counter market must appoint at least two firms to serve as the company's *designated securities companies*. These securities firms will then register the company's shares with the KSDA on behalf of that company. Transactions in the shares will then be handled by the designated securities companies acting as brokers. Settlement is made at the time of acquisition. The trading volume of this market is and will continue to be very small. This reduces investor liquidity and increases share price volatility.

A designated securities company will receive 0.5% of the sales price of an over-the-counter stock as a commission. The authorities require that each designated securities company reports the prices and number of shares which it has traded. The KSDA will then publish the prices and volume of shares which have been traded during the day in the Securities Market Bulletin issued the following day. The KSDA is responsible for making information available to the public concerning the shares registered for sale in the over-the-counter market. It does this by publishing financial statements and other information obtained from the company's designated securities companies. Whether this over-the-counter market will be a success is unclear. However, it is fair to say that the market's development will be hampered by a lack of accurate financial information.

Brokerage fees

Brokerage fees for transactions involving shares listed on the Korean Stock Exchange are as follows:

Value of transaction	Commission rule
Less than W1 million	0.9% of the sale price
W1 million to W5 million	0.8% of the sale price plus W1,000
Over W5 million	0.7% of the sale price plus W6,000

A commission of 2% is charged on the portion of a transaction which is not a multiple of 100 shares. The brokerage commission on bond sales is 0.3% of the sale price. Apart for sales by investment trusts, a securities transaction tax of 0.5% of the sales value will be assessed.

Securities firms

Securities underwriters are authorized under the Securities Transaction Act to issue stocks and debentures on a firm, standby or best-efforts basis. The six largest domestic securities firms in Korea are: Daewoo Securities Co., Daishin Securities Co., Ssangyong Investment & Securities Co., Lucky Securities Co., Dongsuh Securities Co., and Coryo Securities Co. Between them these firms opened 15 overseas offices in Tokyo, New York, London and Hong Kong. Many other smaller securities firms are also doing business in Korea. A list of these smaller firms is set out in the Appendices. Many Korean securities firms have various capital and business cooperation agreements with foreign securities companies. The nature of these agreements is limited because the Korean government prohibits a foreign firm from owning more than a 10% share in a Korean securities firm. Currently, WI Carr and Yamachi each have a 5% equity stake in Daishin, and Nikko has a 5% equity stake in Lucky.

In early 1988, the Korean government allowed the eight largest domestic securities firms to hold up to US$30 million in foreign equities. Insurance firms and investment trusts similarly may hold equities with a value of as much as US$10 million. These firms may use this foreign exchange to gain experience in foreign markets. With some of these funds, Ssangyong Securities underwrote US$2 million in a US$100 floatation of shares in the Asian Development Equity Fund. Lucky Securities underwrote US$1.5 million in bonds with warrants issued by Toyo Wharf & Warehouse. Daishin, Daewoo and Ssangyong participated in the issuance of shares in the US$100 million Thai Fund. However, because the won is appreciating in value so rapidly, substantial amounts are unlikely to be used to purchase foreign stocks and bonds. The government may substantially increase the amount of foreign capital allocated to domestic securities companies in the near future to reduce the current account surplus.

There are twelve foreign securities firms with representative offices in Korea. Since these offices are prohibited from engaging directly in securities-related businesses, they are primarily engaged in market research. While the offices provide no income in the short-term for their head office, they are creating a foundation for the future. Foreign securities offices also provide assistance to Korean companies regarding foreign capital markets and the services and products their head office can provide. Some of the foreign securities firms support the activities of their head offices relating to trading in shares of the closed-end securities funds and on convertible bond issues. As noted previously, foreign securities firms may hold limited equity interests in Korean securities firms.

Foreign securities firms with representative offices in Korea

Company	Country	Opening Date
Nomura Securities Co.	Japan	March 1981
Yamaichi Securities Co.	Japan	March 1982
Daiwa Securities Co.	Japan	December 1982
Nikko Securities Co.	Japan	March 1983
Merrill Lynch Int., Inc.	US	January 1985
Jardine Fleming Co.	Hong Kong	January 1986
Baring Brothers & Co.	UK	August 1986
Vickers da Costa Co.	US	July 1986
Prudential Bache Co.	US	September 1986
WI Carr Sons & Co.	France	September 1986
New Japan Securities Co.	Japan	October 1986
Nippon Kangyo Kakumaru Co.	Japan	October 1986
James Capel	UK	October 1987
Schroder	UK	October 1987
Hoare Govett	US	October 1987

International issues

In December 1981 the US$15 million Korea International Trust (KIT) and the US$15 million Korea Trust (KT) were established. These investment trusts were established and managed by the Korean Investment Trust Co. Ltd and the Daehan Investment Trust Co. Ltd, respectively. Credit Suisse was the lead manager for KIT and Merrill Lynch the lead manager for KT. To meet the demand for beneficiary certificates, each fund issued a second tranche of US$10 million in December 1983 and January 1984 respectively. First Boston and Baring Brothers were lead managers for the KIT second tranche. Merrill Lynch continued as lead manager for KT. As of 1987, the KIT had 2.5 million trust units outstanding. The KT had 1.66 million trust units outstanding. In March 1985, the Korea Growth Trust (KGT), another international investment trust, was established with trusted assets of US$30 million. KGT has 3 million trust units. This fund was issued by the Citizens Investment Trust Co. Ltd and managed by Jardine Fleming. In April 1985, two additional funds came to the market. They were the US$30 million Seoul International Trust (SIT) and the US$30 million Seoul Trust (ST), issued by Korea Investment Trust Co. Ltd and Daehan Investment Trust Co. Ltd, respectively. Baring Brothers & Co. and Vickers da Costa served as lead managers of the former and Prudential Bache the latter. STT and ST both issued 3 million trust units.

Two venture trusts were also established to foster the development of small and medium companies. The Korea Small Companies Trust (KSCT) was established in December 1985. This venture trust has been managed by the Korea Investment Trust Co. Ltd. The other venture trust is the Korea Emerging Companies Trust (KECT), founded in March 1986 and managed by the Daehan Investment Trust Co. Ltd. The KSCT raised US$2 million by issuing 360,000 trust units at US$5.50 per unit. The KECT raised US$3 million by issuing 550,000 trust units at US$5.50 per unit.

As part of its programme to internationalize Korea's securities markets, the government has allowed the formation of two closed-end Korean international funds open to foreign investors. These funds were patterned after the well-established and successful Japan Fund. These funds may purchase shares on the Korean stock exchange as long as total foreign ownership in any Korean company does not exceed 10% or where individual holdings do not exceed 5%. Both funds seek long-term capital growth through investment in the Korean equity market. Annual portfolio turnover of the closed-end funds will not exceed 50%. The first closed-end international fund is the Korea Fund, a US$100 million fund listed on the New York Stock Exchange, managed by Scudder, Stevens and Clark and launched in August 1984. Daewoo Securities Company was the sponsor of this fund and Daewoo Research Institute provided market research for Scudder. The underwriting syndicate included Shearson/AMEX and First Boston.

The second is the Korea-Euro Fund, a US$60 million fund listed in London and underwritten by a 44 member group of European and domestic securities firms. The lead underwriters were Baring Brothers and Ssangyong Investment & Securities. Lucky Securities and five European firms were co-managers of the underwriting. The Korea-Euro fund is managed by Schroder Investment Management Ltd with Dongsuh Securities Co. Ltd and Daishin Securities Co. as co-managers. The Lucky Economic Research Institute provides research for the managers. Shares in the Korea-Euro Fund were heavily over-subscribed even before trading began.

Investing in either the Korea Fund or the Korea-Euro Fund has some inherent risks. First, both are trading at a large premium to net asset value. As the securities market in Korea is liberalized (and foreigners can buy shares directly), this premium will drop sharply. Second, experience has shown that shares in the Korea Fund will drop in value if negative press coverage appears on political or labour issues despite any change in economic fundamentals relating to the fund's underlying investments.

While the domestically managed trusts noted above are difficult to purchase since there are no official market makers, they do trade at a much lower premium to net asset value then the Korea Fund or the Korea-Euro Fund. It has been reported that the leading market makers in the domestic trusts are James Capel, WI Carr and Vickers de Costa. Premiums to net asset value on the domestic trusts range from 80–100%. The broker will typically charge approximately 10% of the premium as his margin in arranging a sale. Estimates put the collective market value of the trusts at approximately US$600 million which is substantially more than the US$145 million initially raised. In general, the domestic trusts have concentrated on blue chip stocks in the manufacturing sector.

The Korea Fund issued a second tranche in May 1986, raising its paid-in capital from US$60 to US$100 million. The Korea-Euro Fund also issued a second tranche of US$30 million in July 1988, raising its capital base to US$60 million. The performance of both funds has been excellent. The Korea Fund had net assets totalling US$213.5 million at the end of 1987, an increase from US$154.1 million a year earlier. Thus, the yearly increase in value for the Korea Fund was 38.5% The assets of the Korea-Euro Fund increased from US$30 million to US$38.3 million during the eight months it was in existence in 1987, which is an increase of 27.6%. The values of the funds have been enhanced recently by the appreciation of the Korean currency since assets are denominated in won and distributions are made in US dollars. Both funds are currently trading at a large premium to actual value. Despite the larger premium, the two closed-end funds are much easier to trade than the seven on-shore trusts, since they are quoted and thus more liquid.

The Korean government has recently announced a plan to form a fund with capital of at least US$30 million during 1988 which will engage in securities trading on foreign markets. Several securities firms will be allowed to manage this foreign stock investment fund. A *Korea Asian Pacific Fund* is also rumoured to be in the planning stages, which might be listed in Tokyo.

The bond market

A number of types of bonds are issued and traded on the Korean bond market. In addition to treasury bills (TB), the Korean government issues Monetary Stabilization Bonds (MSB), Foreign Exchange Equalization Fund Bonds (FEEFB), Grain Bonds, National Housing Bonds and National Railroad Bonds. These government bonds are issued irregularly by the government to further specific government objectives. The bonds generally have relatively short maturities (usually one year or less), and pay interest upon redemption rather than at periodic intervals. These bonds are primarily purchased by financial institutions and are generally held to maturity. A limited secondary market exists for government bonds.

The Korean government issued a large number of bonds in 1987 in exercising its monetary and fiscal policies. The bonds were issued to reduce the money supply given the growth of the economy and the current account surplus. The three main categories of bond issues and their aggregate amounts for 1987 were as follows:

FEEFB	W1.5 trillion
MSB	W9 trillion
TB	W1 trillion

Because of increasing liquidity in financial markets caused by election spending, the Olympics, government social welfare expenditure and the increasing current account surplus, increased bond issuance is very likely during 1988. In particular, offerings of MSBs should increase to as much as W15 trillion, issues of TBs should be W2.5 trillion and FEEFBs, 3 trillion.

Bonds are also issued by local governments, most notably the cities of Seoul and Pusan. These bonds are often purchased by individuals and companies in return for engaging in certain transactions (e.g., purchasing an automobile). Bonds issued by local governments have terms and maturities which are similar to government bonds.

Corporate bonds are the third variety of bond which exists in Korea. These bonds may be issued within limits established under the Korean Commercial Code which consider the issuing company's paid-in capital and reserves. Listed companies are allowed to issue bonds under less rigorous standards than non-listed companies. For a number of reasons, corporate bonds are nearly always issued with a bank guarantee. Interest rates and other terms at issuance are determined by the government. Since the interest rate on corporate bonds may not reflect the market, bonds are usually sold at a discount. Corporate bonds are typically issued by an underwriting syndicate.

Certain Korean banks are allowed to issue bonds secured by their assets. Interest rates and terms of such bonds are regulated by the government. Interest on bank bonds is paid at maturity. Korean branches of foreign banks are not currently allowed to issue bonds.

A prior approval must be obtained if bonds are to be listed on the Korean Stock Exchange. If an issue meets the following requirements, the bonds may be listed:

1. Paid-in capital of the company must exceed W500 million
2. The par value of the bond must exceed W200 million
3. The bonds must have been issued within one year from the date of listing.

Certain debentures issued under special laws and government bonds are exempt from the listing requirements.

As of early 1988, domestic bonds carried yields averaging approximately 12% annually. Relative to the large growth seen in the stock market, the domestic corporate bond market grew more modestly during 1987. The total outstanding amount of listed bonds at the end of 1987 was W26,303 billion. During 1987, W3,123.1 billion was raised in domestic bond issues, an increase of 14% over 1986. Growth in this sector continues to be limited by the relatively small and thinly traded secondary market. Another drag on growth in this market is the unreliability and paucity of information on corporate issuers. During 1988, government bonds issued to control excess liquidity may crowd out corporate issues. Growth in new corporate bond issues should be less than 15%. By early 1988, over 80% of total bond trading involved MSBs.

Certain Korean companies have been authorized to issue straight bonds in international capital markets. Prior government approval is a prerequisite for such issues. Only Korea's strongest companies have been allowed to issue bonds overseas, so that defaults by weaker companies do not diminish Korea's credit standing.

As part of the Korean government's internationalization effort, some Korean companies have been permitted to issue currency denominated bonds including convertible bonds, bonds with warrants and depository receipts in foreign markets. To date, only convertible bonds have been issued. The government has issued rules under which only companies with net assets above W50 billion, whose share prices are above the weighted average and which have a sound financial status may issue convertible bonds on foreign markets.

The convertible bond form has been used in preference to depository receipts and bonds with warrants since it allows companies to make the float at a conversion price which is ahead of the market. In December 1985, Samsung Electronics floated US$20 million of convertible bonds with a 5% coupon on the Euromarket. The lead manager for the Samsung issuance was SG Warburg. The bonds were sold in bearer form on the London market. The conversion price is calculated on the arithmetic mean of the average price of the share immediately prior to the issuance of the convertible bonds and prior to the commencement of the conversion period. Daewoo Heavy Industries issued US$40 million worth of convertible bonds in May 1986, Yukong also floated US$20 million of convertible bonds in June 1986. The lead issuers were Nomura Securities and Goldman Sachs, respectively. Both bond issues carried a 3% coupon. Goldstar completed issuance of US$30 million in convertible bonds on the Euromarket in August 1987.

By the end of 1987, a total of US$110 million in convertible bonds issued by Korean companies had been floated overseas. Conversion of the bonds is dependent upon the issuance of an authorization by the Korean government. For example, the Samsung Convertible bonds issued in 1985 contain a provision allowing conversion on October 19 1987 or on the date the Korean government allows direct investment by foreigners in Korean stock. The conversion period began in November 1987 for the Daewoo issue, January 1988 for the Yukong issue and February for the Goldstar issue. In December 1987, the Ministry of Finance announced that convertible bonds could be converted to Korean shares regardless of whether foreigners were given direct access to the Korean equity market. Since no details regarding the announcement have been issued, no conversions of the bonds have yet taken place. Foreign securities firms continue to press the Ministry of Finance for clarification.

Korean financial circles indicate that Hyundai Engineering & Construction Co., Samsung Semiconductor & Telecommunications Co., Tongyang Nylon Co. and Kia Motors Co. will be the next companies to issue convertible bonds on a foreign market. The Ministry of Finance has announced its continued support of convertible bond issues on foreign mar-

Korean financial circles indicate that Hyundai Engineering & Construction Co., Samsung Semiconductor & Telecommunications Co., Tongyang Nylon Co. and Kia Motors Co. will be the next companies to issue convertible bonds on a foreign market. The Ministry of Finance has announced its continued support of convertible bond issues on foreign markets and has raised the ceiling on international convertible bond floats to US$75 million. Guidelines for the issuance of depository receipts and bonds with attached warrants in international markets can be expected from the Ministry of Finance in 1988 or 1989.

Foreign direct investment

The Korean government has actively promoted direct foreign investment in Korea since the first five-year economic development plan was adopted in 1961. A major step in regulating the flow of foreign investment was the enactment of the Foreign Capital Inducement Law in 1966. While this law has undergone substantial and numerous revisions it continues to be the principal law governing foreign investment. Under the Foreign Capital Inducement Law, the government must approve any direct investment into Korea. The Ministry of Finance has been granted authority to approve or reject foreign investment applications, after consulting with the relevant ministries. In 1984, the Korean government changed its policy by adopting a *negative list* approval system. Under this system, unless an industry is listed by the government as being prohibited or restricted, a foreigner may obtain an approval to make a direct foreign investment in a joint venture or wholly owned subsidiary. Industries are classified on the negative list by utilizing the Korea Standard Industrial Classification System (KSIC). In general, foreigners may invest in nearly all manufacturing-related industries. Most of the prohibited sectors relate to areas where the government wishes to avoid foreign domination or where foreign expertise is not needed (for example, utilities, strategic raw materials, distribution services).

An application for investment in Korea by a foreigner must be submitted to the Ministry of Finance. The form of the application requires that considerable supporting documentation be prepared. The Ministry of Finance will refer the application to relevant ministries which have an interest in the activity in question. The application will also be referred to the Fair Trade Office of the Economic Planning Board if a joint venture is involved and a determination will be made whether the agreement contains provisions which are unfair to the Korean party. Modifications can be (and usually are) required by the Fair Trade Office. In some cases, the equity share which may be held by the foreign party will be restricted.

The Foreign Capital Inducement Law has specific requirements concerning when a decision will be made on an investment depending upon the amount invested, the activity involved, and the foreign equity share in the venture. These rules change often, so the author will not recite the current rules here. Investments approved under the Foreign Capital Inducement Law may also be eligible for certain tax benefits including exemptions from tax on dividends and income, income taxes on expatriate employees and customs duty exemptions and reductions.

In general, foreign direct investment which results in a transfer of technology will be favoured. If the activity in question will involve the exportation of products or if the activity will assist other companies export products, approval will also be more likely. Investments by foreigners which will concentrate on domestic markets will be allowed in more limited cases, and in many cases will not be allowed unless a Korean company is given a substantial equity stake in the venture.

Korea will continue to encourage foreign direct investment in order to reduce the country's dependence upon funds loaned by foreigners. However, as Korea's domestic savings rate increases and the total foreign debt of Korea is reduced by the continuing current account surplus, favourable treatment and special tax and other benefits will gradually be reduced and eventually eliminated.

During the early 1960s, foreign investments were made primarily by US companies. However, with the normalization of diplomatic relations with Japan (which took place in 1965), Japanese companies began to make investments at a rapid pace until they are collectively the largest investors in Korea today. A significant portion of the Japanese investment has been made by Koreans who are long-term residents of Japan.

As a general rule, the investments made by US companies are relatively larger in amount than those made by Japanese investors. US and European investors tend to be large, well-established multinational companies whereas the Japanese investors are smaller both in the size of the investment and the size of the company making the investment.

According to the Ministry of Finance, inflows of foreign investment in Korea in 1987 reached US$1,060 million for 373 cases on an approval basis, an impressive increase of 300% in value and 182% in the number of cases as compared with the preceding year's US$354 million for 205 cases. Japan topped the list with US$494 million for 207 cases followed by the United States with US$255 million for 93 cases, and Europe with US$210 million for 40 cases. Of the foreign investment total, US$208 million was for investment

Foreign investment in 1987

By country (unit: $1 million)

Country	1986 Case	Amount(A)	1987 Case	Amount(B)	Increase rate (B/A:%)
US	48	125	93	225	180
Japan	109	138	207	494	360
Europe	26	63	40	210	330
Other	21	28	33	131	470
Total	205	354	373	1,060	300

By industry (unit: $1 million)

Industry	1986 Case	Amount(A)	1987 Case	Amount(B)	Increase rate (B/A:%)
Electric & electronics	36	67	79	208	310
Machinery	50	91	79	208	230
Chemicals	29	31	45	153	490
Tourist hotels	4	62	8	249	340
Other	86	103	162	142	138
Total	205	354	373	1,060	300

Source: Korea Foreign Trade Association

in 79 electric and electronics industry projects, US$208 million for 79 industrial machinery projects, US$153 million for 45 chemicals industry projects, and US$249 million for eight hotel projects.

Monetary policy

The Bank of Korea is responsible for supervising banking institutions in addition to being the nation's central bank. Monetary policy is made through changes in the terms and conditions of rediscounts, open market operations in securities and in setting reserve requirements. Monetary policy can also be made by altering the terms under which the Bank of Korea makes loans to the government. The Korean government also achieves monetary policy objectives by establishing maximum interest rates, by placing limits on bank fees, and by controlling the issuance of loans by commercial banks.

The following sections describe how monetary policy is made in Korea:

Rediscount policy. The Bank of Korea extends loans to commercial banks and discounts, among other instruments, commercial bills, export trade bills, government-backed bonds, and corporate debentures. A change in the discount rate can force banks to alter their lending rates. The Bank of Korea has used its ability to raise and lower interest rates on loans to financial institutions as a means of influencing monetary policy. The effectiveness of this rediscount policy has been limited by the excess of demand for bank credit and other types of financing. As a result, controls on interest rates and banking fees have been used as a supplement to other monetary policy measures. The most significant component of the rediscount policy of the Korean government has been the establishment by the Bank of Korea of rediscount ratios under which funds are loaned to financial institutions. In addition, rediscount policy can be effective by altering the interest rate charged on temporary loans to financial institutions made to make up deficiencies in reserve funds.

Open market operations. The Bank of Korea engages in buying and selling government-issued securities, government-guaranteed securities and MSBs issued by the Bank of Korea. The Bank of Korea has been hampered in its ability to conduct open market operations by the relatively thin market for such instruments. Most purchases and sales by the Bank of Korea involve other financial institutions. The Korean Monetary Board has authority over the issuance and repurchase of MSBs by the Bank of Korea. In general, interest rates on MSBs have reflected interest rates on time deposits with commercial banks. MSBs are generally issued to financial institutions but are occasionally issued to the non-financial institutions and the general public.

Reserve requirement policy. The Bank of Korea is authorized to establish reserve requirements which banks must maintain for their liabilities to depositors. The amount of reserves required varies depending upon the type of bank involved and the nature of its business. Reserve requirement ratios in Korea have been historically high by international standards. Owing to limitations in the effectiveness of the previous two monetary policy techniques, reserve requirements are heavily relied upon by Korean monetary authorities. The Bank of Korea also has authority to require banks to place funds in an interest-bearing stabilization account.

Credit controls. The Korean government, through its direct and indirect control over the financial institutions, has been able to influence monetary policy by placing ceilings on total loans and by approving individual loan applications. For example, in January 1988 the Economic Planning Board announced that large businesses would not be eligible for new loans unless cumulative outstanding borrowings were less than levels at the end of 1987. The Korean government can also require that loans be made to certain sectors (e.g., small businesses) or to certain industrial sectors (e.g., high technology). The fact that such a small number of companies dominate the economy of Korea makes it much easier for the government to make monetary policy in this manner.

A major component of Korean monetary policy since the first five-year plan has been the *cheap money* policy. To implement this policy, the government set interest rates on savings and loans at artificially low levels. Until recently, high inflation rates acting in concert with the cheap money policy have resulted in a negative real interest rate on both deposits and loans. This resulted in an excessive demand for funds as Korean companies borrowed for speculative as well as purely economic reasons. Since demand for funds exceeded supply, government rationing was necessary. It is through this rationing process that the government was able to force businesses to comply with its policies and to grant favours to companies which supported the political party in power. This cheap money policy had the negative consequence of discouraging domestic savings and increasing foreign borrowings and increasing inflation.

The cheap money policy also had the effect of producing high inflation levels.

The battle against inflation in Korea has been successful in recent years through tight monetary and fiscal policies which have resulted, among other things, in a reduction of annual money supply (M2) growth from 27% in 1981 to 18.8% in 1987. Other anti-inflation policy measures have included selective price controls, tax reductions and restrictions on real estate speculation. Korea's tight monetary policies will continue for the foreseeable future as the Roh administration has stated that keeping the growth of the money supply (M2) below 18% is a major priority. This will be a very difficult standard to meet.

Annual consumer price increases in Korea 1970–87			
1970:	15.9%	1979:	18.3%
1971:	13.5%	1980:	28.7%
1972:	11.7%	1981:	21.3%
1973:	3.1%	1982:	7.2%
1974:	24.3%	1983:	3.4%
1975:	25.3%	1984:	2.3%
1976:	15.3%	1985:	2.5%
1977:	10.1%	1986:	2.3%
1978:	14.4%	1987:	5.8%

The inflation level for 1987 was particularly disappointing for Korea's economic planners. The consumer inflation rate increase was 5.8% in 1987. The primary contributors to this increase were rises in food, housing, tuition and transportation costs. Wholesale prices rose by 1.7% during 1987 despite government-ordered reductions in energy prices.

The money supply (M2) in Korea grew 17.3% in 1986 and 22.5% in 1987. This growth substantially exceeded the government's announced target of 18%. Money supply growth should again exceed 20% in 1988 despite a Herculean effort by Korea's monetary authorities. Experts predict that approximately W1.4 trillion of MSBs will be issued during 1988.

Offshore debt financing

The rapid growth of the Korean economy has been accompanied by a correlative increase in the level of foreign debt. In order to achieve the 8% average growth in GNP from 1962-85, the government estimated that the Korean percentage of investment to GNP needed to be 29.2% per year. For a number of reasons (including high inflation, capital flight and low per capita income), the domestic savings rates did not meet the required levels. The domestic actual savings ratio averaged 24.1% from 1962-85. Other real savings rates during this period were as follows:

Korea's investment and savings (averages)			
Period	Gross domestic investment	Domestic savings	Foreign savings
1967–71	26.3	14.8	10.5
1976–76	26.1	19.3	7.4
1977–81	31.0	25.5	5.4
1982–85	30.4	25.6	4.7

Source: Bank of Korea, Annual Economic Statistics

Seoul – the centre of political, educational and economic opportunity since it was established by the Yi Dynasty in 1392: and the home of over 11 million of Korea's 42 million population

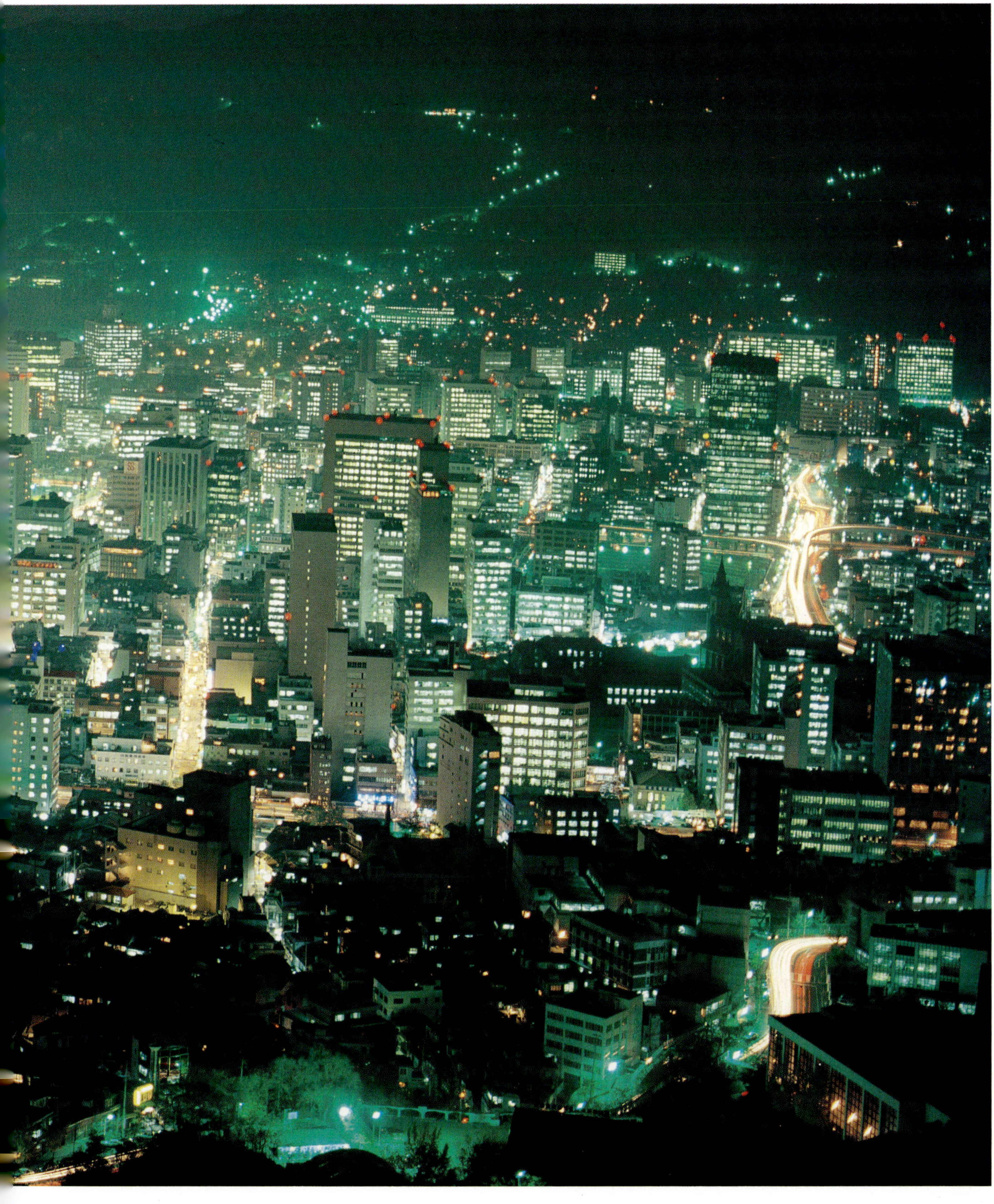

Continual borrowing from offshore sources by Korea during the 1960s, 1970s and early 1980s resulted in a total foreign debt level of US$46.7 billion in 1985. Since that time, Korea's current account surplus and rising domestic savings rate has allowed the country to reduce its foreign debt to approximately US$35 billion by the end of 1987. Foreign debt should fall below US$30 million by the end of 1988.

There are three types of offshore debt financing in Korea. The first type is a *public loan*, which is what its name implies. These loans are typically made by a foreign government or an international economic development organization, such as the World Bank or the Asian Development Bank. Loans of this type must be approved in advance by the Ministry of Finance. Public loans are usually tied to a specific development project and have a rate of interest below that available to commercial borrowers.

The second type of foreign borrowing is made by *foreign exchange banks* to raise funds on a long-term basis. Borrowers of this type include the Korean Development Bank, the Korean Long Term Credit Bank and commercial and merchant banks. Any such foreign borrowing by a bank must be approved in advance by the Ministry of Finance pursuant to the Foreign Exchange Control Law and Regulations.

The third variety of foreign borrowing will be extremely limited in 1988 since the Korean government is striving to reduce liquidity and the nation's foreign debt. If a loan has a term longer than three years and an amount in excess of US$1,000,000 (a *long-term loan*), it must be approved by the Ministry of Finance. The relevant statute under which an approval may be granted is the Foreign Capital Inducement Law. Owing to its high liquidity, as of 1988 only loans made to retire expensive debt or loans made to strategic industries will be approved.

Loans which are not long-term under the above criteria must be approved in advance by the Bank of Korea under the Foreign Exchange Control Law and Regulations. Loans of this type are available only in limited situations, such as raw materials imports or overseas construction. The terms of any such loan must meet strict requirements established by the Bank of Korea.

Public finance

The Korean government plans and administers the nation's budget through what it calls a *unified budget system*. The cumulative amount of this budget in 1986 was W24,907.6 billion (excluding intra-government transactions). The budget amount was equal to 31% of GNP during that year. In order to eliminate waste and increase efficiency, the Korean government adopted a zero-based budgeting system beginning in 1983. The government also instituted a budget freeze in 1984 in order to keep public sector spending in check.

The public sector in Korea comprises central government, local governments and government owned and affiliated companies and institutions. While the majority of government expenditures fall within the formal budget, a large number of extra budgetary accounts must also be considered in examining public finance in Korea. The central government maintains a general account, 17 special accounts and 25 government management funds. In addition, general and special accounts are maintained for special cities, provinces as well as local governments. Government owned businesses and enterprises are organized into five sectors: railways, telecommunications, monopoly, procurement and grain management. Budgets for government owned companies are organized on this basis. A Grain Management Fund and a Procurement Fund are also maintained by the government.

The budget formulation process originates with officials at the Economic Planning Board when they propose guidelines on the coming year's budget to all government ministries and organizations. Budget estimates are then submitted by the ministries and organizations to the Budget Bureau of the Economic Planning Board for review. Preliminary budget figures are then submitted to the Council of Economic Ministers for consideration. The Budget is then approved by the President if the figures are acceptable. A Budget Bill is then forwarded by the President to the National Assembly for approval. As a rule, the National Assembly receives the Budget Bill in late September. The National Assembly, through the Special Committee on Budget and Accounts, then reviews the budget and holds hearings on general and specific expenditure levels. The Budget must be enacted prior to December 1 so that preparations for the new fiscal year (a calendar year) can be started well before it begins.

The Ministry of Finance implements the budget by distributing the respective shares of the budget to the relevant ministries and organizations on a quarterly basis. The Board of Audit and Inspection, which operates under the responsibility of the Blue House, is responsible for auditing the expenditures made by the government. This board submits a report to the Minister of Finance each year on the government's expenditure during the previous year.

On the revenue side of the public finance equation, the Korean taxation system is made up of both national taxes and local taxes. National taxes are: internal taxes, customs duties, defence taxes and education taxes. Local taxes are divided into ordinary taxes and earmarked taxes. Tax revenues are supplemented by profits from government-run monopolies (such as cigarettes and ginseng). These non-tax revenues are not a major factor since over 95% of Korean government revenues are derived from taxes. The amount of taxes collected by the government has risen substantially as Korea has grown. Adding to the revenue gain has been increased tax enforcement efforts. Between 1960 and 1980, the tax ratio in Korea increased from 11.1% to 17.2%. The ratio of tax burden to GNP will increase from 17.9% in 1988 to 19.2% in 1991. While the share of revenue supplied by customs taxes will decrease, defence and internal taxes will rise, taking total tax receipts in 1988 to W18,919 billion. In 1988, the income tax rate ranges from 6–55%, but before 1990 the top rate should drop to 50%, the basic exempted amount should increase and the number of marginal rates should be reduced.

Because of pressing economic needs, the Korean government has recorded chronic budgetary deficits since the 1960s. Efforts are now being made to balance the nation's budget. Because of the recent current account surplus, the central government is no longer borrowing new funds abroad to financing its activities. In general, the government has increasingly devoted budgetary expenditure to improving public welfare and promoting economic stability rather than expanding national investment and subsidizing private industry. This trend should accelerate in the years to come.

Taxation

Taxes are imposed in Korea by national, provincial and local governments. The Office of National Tax Administration (ONTA) is responsible for assessing and collecting taxes at a national level. The following taxes are assessed by the ONTA:

- corporate income tax
- individual income tax
- defence tax
- value added tax
- other taxes including consumption taxes, securities transaction tax, stamp tax, education tax, customs duties and excise taxes, telephone tax and liquor tax.

Provincial and local governments impose inhabitant taxes, acquisition taxes, registration taxes, licence taxes, property taxes, city planning taxes, community facility taxes, workshop taxes and farmland taxes.

A progressive rate scale is imposed upon both corporate and individual income taxes based upon a calculation which allows for certain deductions and exemptions. Korea imposes its graduated income tax on all businesses which have a *domestic place of business* in Korea (called a *permanent establishment* if a bilateral tax treaty rather than the corporation tax law is applicable). This determination is very complex regardless of whether it is made under a tax treaty or a Korean statute. Korea has entered into tax treaties with Belgium, Canada, Denmark, France, Japan, Thailand, the UK, West Germany and the US, among other countries. Non-resident foreign companies without a domestic place of business are generally taxed on gross receipts derived from Korean sources. Non-residents are not subject to defence tax, but may not claim exemptons or deductions. If a foreign company does not have a domestic place of business, Korea will assess taxes on Korean source income through withholding. Withholding obligations are imposed on interest and incidental income, salaries and wages, retirement benefits, dividends, professional income, gain from the sale of shares, gross revenue and royalties.

Koreans and foreigners who are residents of Korea for tax purposes are subject to income tax on their worldwide income. Branches of foreign companies are treated as a separate taxable entity. While the determination of whether an individual is a resident is made based upon the facts and circumstances in each case, in general, a resident is a person who is domiciled or has been a resident in Korea for more than one year. A defence surtax on income tax is also imposed. Certain deductions and personal exemptions are allowed against income tax on residents.

Korea imposes a value added tax (VAT) on the sale of goods and services in Korea, as well as the importation of goods into Korea. The current VAT is 10%. The tax is assessed on the sales price of the goods or service with a credit being given for VAT paid when the raw materials or goods were purchased. Foreign companies need not collect VAT unless they have a place of business in Korea since the importer must pay the VAT in such a situation. While exported goods are not subject to VAT, a return may nevertheless need to be filed.

The Korean government has created a range of tax incentives and concessions to encourage certain activities which further policy objectives. Certain investments made by foreigners can qualify for a tax holiday, customs duty exemptions and tax payment deferrals. Licence or royalty payments and salaries paid to foreign managers and technicians can also be exempted from taxation. Other tax benefits granted to foreigners include special depreciation formulas and tax exemptions on certain types of interest. A detailed description of these benefits is beyond the scope of this book. Furthermore, the government is gradually phasing out these benefits as the need for the incentives is reduced. Rules cited here would quickly be out of date.

The Korean tax system is extremely complex and substantial penalties may be imposed if a company or individual fails to make required filings or pay taxes which are due. Certain deductions which are available abroad are not available in Korea (and vice versa). For this reason, proper tax advice is essential in structuring any business transaction involving the Republic of Korea.

Foreign exchange controls

A number of Korean laws and regulations govern foreign exchange transactions. These include the Foreign Exchange Control Law, the Enforcement Decree of the Foreign Exchange Control Law, the Foreign Exchange Control Regulations, The Foreign Capital Inducement Law and the Foreign Trade Law. Failure to comply with Korea's foreign

Construction of a new furnace at the Pohang Steel Works; one third of Korea's steel output is typically exported to the US and Japan

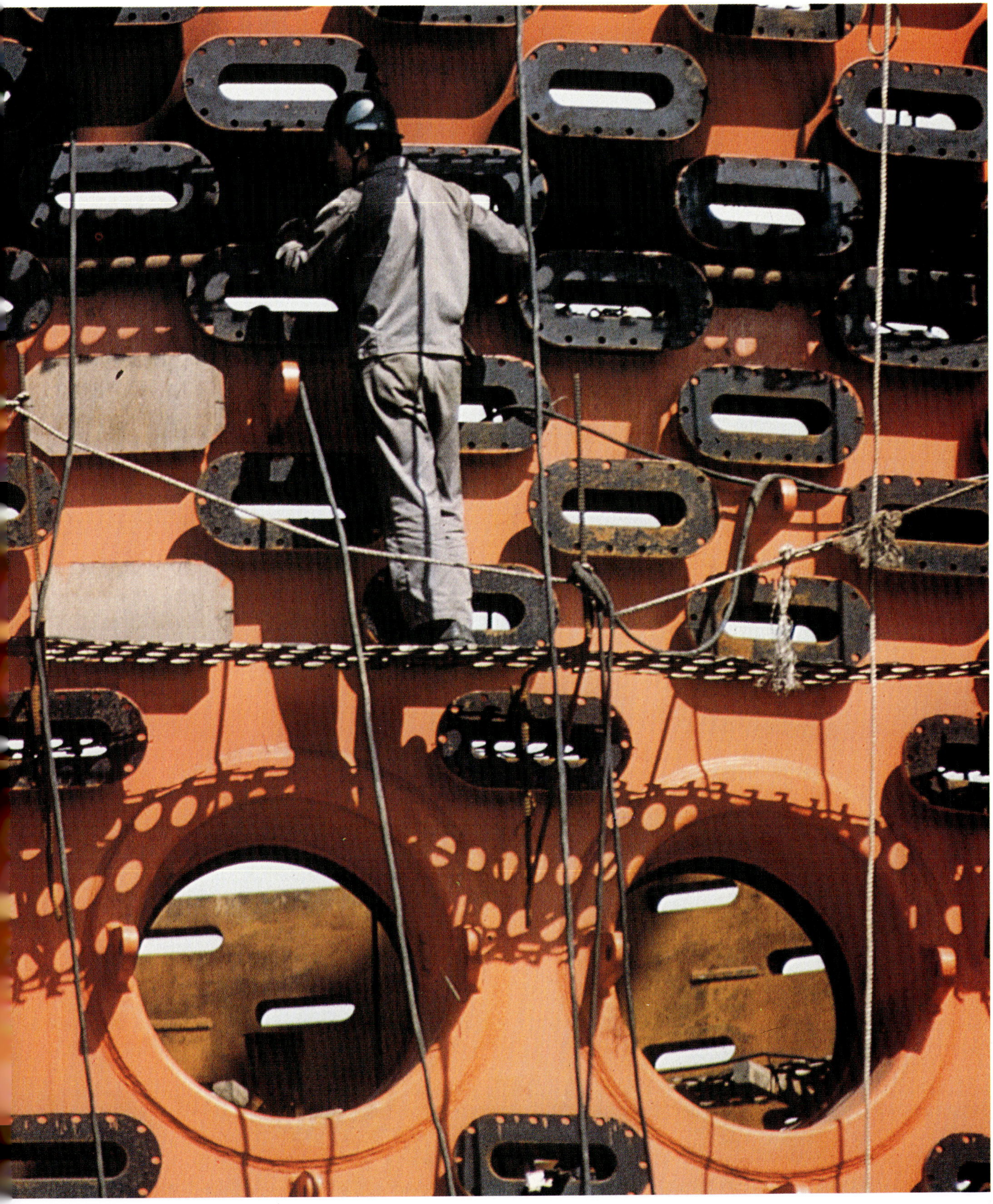

exchange control laws may or may not invalidate the legality of a transaction but will certainly affect the enforceability of contractual rights. Furthermore, violations of such laws may subject the offender to heavy fines, forfeitures and prison terms. In some cases, a violation of a foreign exchange control law may be a capital offence.

The supreme administrative body with authority over foreign exchange transactions in Korea is the *Foreign Exchange Deliberation Committee*. This committee is composed of the Minister of Finance, the Minister of the Economic Planning Board, the Minister of Agriculture and Fisheries, the Minister of Trade and Industry, the Governor of the Bank of Korea and two individuals appointed by the President of Korea. This committee reviews and approves the actions of the Ministry of Finance. The Ministry of Finance is responsible for establishing and monitoring the activities of foreign exchange banks and for the preparation of the *Foreign Exchange Supply and Demand Plan*. This last function is extremely important as the plan establishes basic foreign exchange policies for each fiscal year. The Bank of Korea has been authorized by the Ministry of Finance to perform certain functions in relation to foreign exchange including the determination of residential status and the establishment of the foreign exchange concentration rate.

Day-to-day implementation of foreign exchange transactions has been delegated by the Bank of Korea to authorized *Foreign Exchange Banks*. These banks are classified as either Class A or Class B foreign exchange banks. Class A foreign exchange banks are authorized to conduct normal foreign exchange activities including the holding of foreign currencies in foreign countries, the purchase and the sale of foreign exchange. Currently, all city banks, foreign bank branches, six merchant banks and most local banks are Class A foreign exchange banks. Class B foreign exchange banks may only engage in limited foreign exchange activities.

Foreign exchange banks are required to maintain a continuously over-bought position to an extent set by the Bank of Korea. If a bank holds foreign exchange in an amount which exceeds a specified limit, such currency must be sold to another bank. Thus, an interbank market exists in foreign exchange. This market is supervised directly by the Bank of Korea (trading takes place on Bank of Korea premises). No brokers are involved in the interbank market but some banks act as the agent of others. The Bank of Korea becomes directly involved in this interbank market only if a bank cannot reduce its over-bought position.

Class A foreign exchange banks are authorized to issue payment guarantees in foreign currency to residents and non-residents for specified transactions at the request of both residents and non-residents as well as foreign banks. Such guarantees will be issued only if the underlying agreement is approved in advance. Retroactive approvals will not be granted. Foreign exchange banks may also make loans of foreign currency (within a lending limit set by the Ministry of Finance). These loans can be made only under circumstances and terms established by the Bank of Korea.

Forward foreign exchange transactions have been allowed in Korea only since 1980. This is the closest equivalent to hedging currency risk in Korea. At present, forward foreign exchange transactions occur in the US dollar, the Japanese yen, the deutschemark and the pound sterling. The biggest change in the foreign currency area came in October 1987 when Class A foreign exchange banks were given permission to offer foreign exchange products, such as futures, options and foreign currency and interest rate swaps. However, these futures transactions may only involve currencies other than the won. This limitation will severely limit the size of the foreign exchange market. Trading in this area should be more profitable for foreign banks than for domestic banks, given their greater experience in international markets.

Any transaction involving the flow of currency or creating a credit relationship between a resident of Korea and a non-resident requires approval of the foreign exchange control authorities under the Foreign Exchange Control Law and the Foreign Exchange Control Regulations. Domestic Korean corporations (including those owned by foreigners) and Korean branch and liaison offices of foreign corporations are residents for foreign exchange control purposes. A foreign national is deemed a resident of Korea for foreign exchange control purposes if he works at a place of business in Korea, engages in business in Korea, or maintains a domicile or residence in Korea for six months or more. Two types of approvals are generally required for foreign exchange transactions under the Foreign Exchange Control Regulations: (i) approval of the underlying agreement and (ii) sanction of specific payments made under authority of the underlying agreement. In most cases, if approval of the underlying agreement is not obtained in advance, authorization for specific payments will not be given.

Since the won is not freely convertible into foreign currencies, visitors must retain all foreign currency exchange receipts to allow reconversion upon departure from Korea. If exchange receipts are not kept, the re-exchange limit is US$100 for each visit. The receipt of foreign exchange into Korea is freely permitted, while payments of foreign exchange is strictly controlled. In principle, all foreign currency brought into Korea by anyone other than a short-term visitor must be converted into won or deposited in a special account, such as a resident or non-resident bank account. In practice, the recent current account surplus has caused the government to permit certain types of foreign currency accounts. Travellers may carry a reasonable quantity of cash of a foreign currency. Many merchants in Seoul (particularly in the Itaewon shopping district) will quote prices and accept payment in US dollars. While technically illegal, this practice is tolerated by the Korean government. An active US currency black market operates in a surprisingly sophisticated fashion. The brokers make a commission on the spread between official and black market rates, paying a higher rate for $100 bills.

All transactions between Korea and other countries must be implemented in one of the 53 designated currencies prescribed by the Minister of Finance on the basis of their convertibility, transferability, and other factors. The exchange rates of the following 22 currencies are quoted daily: US dollar, pound sterling, deutschemark, Canadian dollar, Italian lira, French franc, Hong Kong dollar, Australian dollar, Dutch guilder, Danish krone, Austrian schilling, Swiss franc, Swedish krone, Belgian franc, Norwegian krone, Japanese yen, Saudi Arabian riyal, Kuwaiti dinar, Bahraini dinar, UAE dirham, Singapore dollar, Malaysian ringgit. The most widely accepted currencies in contracts and commercial transactions are the US dollar, the Japanese yen and the deutschemark.

Some gradual liberalization of Korea's foreign exchange laws can be expected as the Korean government struggles to deal with its unprecedented trade surplus. For example, Koreans are now allowed to take US$10,000 when they travel overseas, twice the amount formerly allowed. Koreans moving to foreign contries as immigrants are now allowed to take US$200,000, again twice the previous ceiling. In the future, Korean securities firms should be given greater amounts of foreign exchange to invest in foreign markets. Korean businesses which are developing markets in developing and third world countries should also be given access to increased amounts of foreign currency.

Chapter X
Ground rules for doing business

Choosing your business partner

The importance of choosing a Korean business partner or employee with the proper business and political connections (which arise from his relationships with key persons) cannot be over-emphasized. A person's network of relationships literally define his life. For this reason, doing business in Korea is an intensely personal process. Whether a Korean businessman is from a good (i.e., upper class) family, attended a prestigious university or high school, is from the right home town, is married to the right person or is the proper age can mean the difference between success or failure.

The importance of proper business connections means that it can be extremely useful to involve a consultant at an early stage. This is particularly true if the foreign company is doing business for the first time. This individual's relationship with and connections to officials in government, financial institutions, and private business can immeasurably increase the chance that the enterprise will be profitable. Because of the difficulty in finding a suitable consultant, it is generally advisable to retain a foreign consultant who has lived and worked in Korea for many years. This person can not only provide the proper connections, but can serve as a cultural translator as the business deal evolves.

Obtaining a proper introduction may be the most important action a foreigner can take in Korea. Most foreign businessmen find that an introduction is essential in order to obtain appointments with high level Korean businessmen. As will be discussed later, the higher one starts in the Korean business organization, the easier it is to obtain a favourable decision.

The respect a Korean has for foreigners (especially in the early stages) is directly related to the respect he has for the intermediary. If possible, the introduction should come from someone whose position is comparable (preferably higher) to the person you desire to meet. An intermediary can also be useful later as a mediator in resolving deadlocks in a negotiation or a business dispute. It is no small matter for a person to agree to give an introduction as he will in effect take a measure of responsibility for the actions of both parties, in the future. For this reason, the decision to provide an introduction is made with a great deal of consideration. It is preferable for the intermediary to personally accompany you to the initial meeting. His presence will be useful in assisting the flow of small talk which should proceed substantive negotiations.

It will never be appropriate for a foreigner to contact a Korean executive for the first time by telephone without using an intermediary. Cold call contacts make Koreans extremely uncomfortable in nearly all cases and can seriously impede a foreigner's ability to build a relationship.

Bankers, lawyers, accountants, diplomats and businessmen from your own country are often able to supply the necessary introduction. Former and current Korean business partners are also an excellent source of introductions. For this reason, it is wise to maintain relationships with individuals who may be in a position to provide an introduction by keeping in periodic contact. Foreign businessmen should adopt the Korean practice of paying short courtesy visits when they are in Korea to keep relationships from becoming dormant even if there is nothing specific to discuss. Sending Christmas or New Year's cards to Koreans you have met is a helpful practice to adopt. Korean businessmen take the sending of such cards seriously and may send hundreds around the world every year.

Business cards

The second step in building a relationship is to explain precisely who you and your company are to the Koreans involved in the transaction. Until a person or entity is known, a relationship is simply not possible. Business cards are essential. The business card identifies your position in the company and thus your relative status. As any foreigner with prior experience in Korea knows, relative status must be established before discussion can proceed. For this reason, do not be offended if the Korean businessman is more interested in your position than in you.

Many foreigners find it advantageous to restructure or re-phrase their title to facilitate doing business in Korea. For example, *Director of Asian Operations* implies more authority than *Assistant Vice-President*. Obviously, restructuring your title is intended to clarify your responsibility and relative level in the organization, not to mislead. In general, the title vice-president has been debased in Korea and implies little about actual status.

The business card should be printed on high quality paper and be of standard size so that it can fit in the catalogue of cards maintained by his secretary. Having your business card duplicated in the Korean language on the reverse side is not necessary but does indicate a certain degree of sophistication regarding the Korean market.

A few items of protocol concerning business cards bear noting. First, the visitor offers his card after the bow/handshake. Second, always offer and accept business cards with two hands rather than one. The left hand should support the right at mid-forearm in this two-handed presentation (vice-versa for left handers). Third, hold the card English side up so it can be read easily. After receiving a business card, it is polite to take a few seconds to read the card. If particularly impressed by your card the Korean may bow again to indicate respect for your position or your company. It is then customary to place the business cards on the conference table arranged in the order of seating for use as a reference during the meeting. Treat the card itself with respect.

Company brochures and annual reports are very helpful in establishing the foreign company's identity. In Korea, bigger is generally equated with better. An executive of a large company has more status and prestige than his counterpart in a small company. High quality production, printing and paper are a good idea since Koreans place a high value on the quality of written materials. When meeting very senior business leaders or government officials it may be necessary to send a copy of your resume prior to the meeting. Bring a copy of your resume to Korea even if you do not expect to need it.

Social relationships

The next step in building a business relationship with the Korean company is to form a personal relationship with your Korean counterparts. Informal social meetings should be scheduled as frequently as possible as part of this process. Expect to be entertained and once entertained be sure to reciprocate. The entertainment does not need to be lavish and should be informal and as intimate as possible. Late night eating and drinking is an integral part of the Korean business world. Men typically meet with friends after work for a meal, drinks and entertainment. Because Koreans are more relaxed while drinking, social drinking provides a unique opportunity for a foreigner to establish close relations. Invitations to play golf or tennis are also useful in building good relations with Koreans, particularly for foreigners who do not drink.

A sincere effort should be made to learn as much as possible about the Korean company and the employees involved in the business transaction or relationship. It is polite to ask Korean businessmen about the schools they attended, the places they have visited overseas, their hobbies and their children, but not domestic political issues. However, Koreans will be interested in political and economic developments in your country.

The giving of gifts is a much more significant part of business etiquette in Korea than in Western countries. Giving small gifts, especially after your return to Korea, is an appropriate gesture. If possible, products made by your company or companies in your region of the country are an excellent choice. Foreign liquor (particularly brand name Scotch, bourbon, white wine and Cognac) and pen sets are always well received. Items with prestigious designer brands are also good choices.

A foreign businesswoman faces particularly difficult problems in building a relationship with Korean men. Business in Korea is and will continue for the foreseeable future to be a male-dominated activity. Many foreign businesswomen have been very successful in developing a good relationship with Koreans, but they have had to work much harder to obtain this result. The typical response of a Korean man when confronted by a foreign woman is to retreat into formal and ritualistic responses. The foreign businesswoman must break through this formal facade by placing the Korean businessman at ease.

Many foreign companies fail to fully take advantage of the opportunity for advancing a relationship which arises when a Korean businessman visits their country. Koreans may feel insecure when travelling abroad and will appreciate your efforts to make their visit less stressful and more enjoyable. Many Koreans do not drive or are unfamiliar with driving abroad and will appreciate your arranging transportation for them.

Many foreign companies also make the mistake of not scheduling sufficient time for a business trip to Korea. Short trips to Korea are seldom useful since they do not allow sufficient time to build the necessary relationships. A Korean businessman will feel that a proper business relationship cannot be developed unless a significant amount of time is devoted by the parties to learning about each other.

It is often said that a Korean negotiates a relationship rather than a contract. The existence of mutual trust and understanding which should arise from this relationship is of primary importance to the Korean businessman. If the business relationship between the parties has been properly nurtured, the foreign businessman may be able to avoid the disputes and litigation often encountered in doing business in other countries. The emphasis on relationship creates some unique problems for foreigners, particularly with regard to contracts.

When business negotiations begin, it is best to talk informally about yourself and your company. It will seldom be productive to immediately begin substantive negotiations. Such directness and haste will be considered impolite and will make a Korean uncomfortable and wary. In most cases, only general information concerning the companies will be exchanged during the initial meeting.

While the foreign businessman should strive to develop the best relationship possible with his Korean counterpart, he must remember that he will always be considered to be an outsider by a Korean. The relationships which are really

Korea's largest chaebol

Group		Major business sectors
(1)	Hyundai Chairman: Chung Se Yung	Automobiles, electronics, construction and engineering, shipbuilding, trading. (32 affiliates).
(2)	Samsung Chairman: Lee Kun Hee	Electronics, semiconductors, trading, textiles. (35 affiliates).
(3)	Lucky-Goldstar Chairman: Koo Cha Kyung	Semiconductors, telecommunications, electronics, petrochemicals, trading. (57 affiliates).
(4)	Daewoo Chairman: Kim Woo Chung	Electronics, automobiles, trading, heavy industries, shipbuilding. (29 affiliates).
(5)	Sunkyong Chairman: Choi Jong Hyan	Trading, oil exploration, petrochemicals, audio and video tapes. (18 affiliates).
(6)	Ssangyong Chairman: Kim Suk Won	Heavy industries, automobiles, cement. (22 affiliates).
(7)	Korea Explosive Chairman: Kim Seung Youn	Petrochemicals, auto parts, oil refining. (21 affiliates).
(8)	Hyosung Chairman: Cho Suk Rai	Textile, petrochemicals, trading. (15 affiliates).
(9)	Hanjin Chairman: Cho Choong Hoon	Air and marine transportation, stevedoring, construction. (13 affiliates).
(10)	Lotte Chairman: Shin Kyuk Ho	Market distribution, tourism, confectionary, construction. (31 affiliates).
(11)	Dong Ah Construction.	(16 affiliates).
(12)	Doosan	Foodstuffs. (21 affiliates).
(13)	Kolon	Textiles, chemicals. (17 affiliates).
(14)	Hanil Synthetic	Synthetic fibres, footwear. (11 affiliates).
(15)	Kumho	Petrochemicals, airline, tires. (19 affiliates).
(16)	Dongbu	Steel. (12 affiliates).
(17)	Kia	Automobiles. (9 affiliates).
(18)	Sammi	Special steels, trading. (7 affiliates).
(19)	Daelim	Construction. (13 affiliates).
(20)	Haitai	Food and beverages. (13 affiliates).

important in Korea (for example, family member, high school or college classmate) will never exist between a Korean and a foreigner. Hiring a Korean-American as an employee will not solve this problem since he only has his race and language skills in common with the Koreans. As an outsider, the foreign businessman must always be careful to protect his interests.

Foreign businessmen may disagree about many aspects of doing business in Korea. However, there is unanimous agreement among foreigners that the choice of the Korean company or companies which you do business with is the single greatest factor in determining whether the end result will be profitable. As a result, this choice should be made carefully after considering a number of candidates.

The chaebol

Approximately 50 Korean business conglomerates (known as chaebol) dominate the Korean economy to an extent difficult for most foreign businessmen to comprehend. The largest five chaebol are in fact super-chaebol and have an enormous impact on the business environment in Korea. The gross revenue of the Hyundai group accounts for nearly 5% of the country's GNP. The 20 largest chaebol are responsible for producing over half the value added in the manufacturing industry. The fact that each of the largest five chaebol had total sales in 1987 which were larger than the combined sales of Taiwan's ten largest companies gives an indication of how these companies dominate the business environment. Relative positions will vary from year to year and will depend upon whether the ranking is based on earnings, turnover or assets.

The dominance of the chaebol is particularly pronounced in export markets. In an effort to increase exports, the Korean government in 1975 created a system for designating *general trading companies*. Each of the major chaebol was encouraged to establish a general trading company. To date, 11 companies have been given general trading company status by the government, including a company that specializes in assisting small companies and which is not affiliated with any of the chaebol. These companies are Daewoo, Hyosung, Korea Trading International, Kukje, Lucky Goldstar International, Kumho, Koryo, Samsung Co., Ltd., Hyundai Corp., Ssangyong and Sungkyong, Ltd. The general trading company system was established to increase economies of scale in the promotion of exports, to provide a mechanism for small companies to export goods, and to shift some of the expense of export promotion from the government to the chaebol.

Approximately 50% of Korean exports were sold by general trading companies in 1988 (approximately US$15 billion of US$30 billion in total exports). The general trading companies were responsible for US$20 billion of US$60 billion of total Korean trading volume in 1986. Average return on sales in 1984 was 0.3%. Approximately 300 overseas branches were maintained by general trading companies in 1985.

The temptation to equate Korean general trading companies with their Japanese cousins (the *sogo shosha*) can lead to erroneous conclusions. Japanese trading companies are heavily involved in structuring and arranging domestic transactions between group companies. This domestic intermediary role increases the ability of Japanese companies to produce goods with the greatest value added. Only after the domestic trading function is satisfied does the Japanese trading company attempt to move the product into overseas markets. In other words, in Japan the trading company creates the group by developing its domestic trading function. In Korea, the group exists already and the general trading company merely facilitates international sales. Korean general trading companies are not allowed to import a wide range of raw materials as this function is assigned to industrial and end-user associations organized by the government. This substantially reduces the role of Korean general trading companies.

While similar to Japanese zaibatsu, chaebol have some unique characteristics. Chaebol are generally still controlled by the entrepreneurs who initially organized the company group. This is true for nearly all Korean companies since approximately 65% of Korean corporations are controlled by the founding individual. Even if the founder has died or relinquished operating authority over the chaebol, a family member (often his eldest son) is usually in firm control of the business. The founding entrepreneur and his family will own all but a nominal portion of the stock of the companies which comprise the chaebol. The shareholders of the chaebol are reluctant to have member companies go public since to do so would endanger the family's control over the business and reveal financial information to competitors. Only significant government pressure combined with tax and financial benefits has resulted in a few of the chaebol companies issuing shares to the public. However, even companies in the chaebol which have gone public have offered only a small portion of their shares to outsiders. Furthermore, publicly held shares may be actually held by members of the family through nominees. The chaebol rely heavily on bank financing rather than the securities markets in order to remain privately held.

A chaebol is composed of a varying number of affiliated companies which are related through an incredibly complex and convoluted scheme of cross-ownership and financings. For example, Hyundai, Daewoo and Lucky-Goldstar each has over 30 related companies in the company group. One important point to note is that the affiliated companies in the chaebol are not typically organized in the parent-subsidiary chain common in western countries. Instead, common shareholders and cross shareholdings generally link the companies together.

Each chaebol has an extended group of sub-contractors, vendors and suppliers with which it deals. Unless the sub-contractor, supplier or vendor is unusually strong, it will be difficult to make sales to more than one primary business group. In effect, these suppliers and vendors are captives of the chaebol. The bond between chaebol and their suppliers and sub-contractors is feudal in nature. The smaller companies are like lords who owe allegiance to a king. No king owes any duty to the lord of another king and vice versa. Once the relationship is severed, the sub-contractor becomes a rogue and may have trouble finding a new chaebol alliance. In practice, the vendors, suppliers and sub-contractors are subjected to many unfair practices such as the withholding of payments, unilateral price changes and the return of goods for credit. Chaebol have been known to place large orders with a sub-contractor, delay payment or reject a large shipment due to alleged defects and then purchase the financially distressed company. The Korean government is stamping out these abuses.

Significantly hindering the development of Korean sub-contractors has been the fact that they have not been able to obtain enforceable long-term supply commitments from the chaebol. Because the sub-contractors are so dependent upon a single chaebol and may have their relationship severed at any time, they are reluctant to invest in the facilities and research necessary to produce high quality technologically advanced products. Sub-contractors also diversify to reduce risk, thus limiting the gains in productivity from specialization.

One major difference between Korean and Japanese business groups is that zaibatsu are much more integrated in terms of industrial structure. Products made by a Japanese company are composed of products made by affiliated companies. In contrast, Korean chaebol are more of a collection of unrelated businesses with common ownership. As a result, the chaebol are much more dependent upon outside sub-contractors for supplies. Since Korean sub-contractors tend to be poorly managed and undercapitalized, their technological capacity is very low. Thus, the chaebol are dependent upon foreign companies for key parts and components.

Most chaebol are striving to increase their degree of vertical integration, based on the need to maintain control and reduce risk. Theoretically, each company in the group should be able to rely on the services of other group companies (e.g. a construction company, an advertising firm, a research institute, a transportation company or a financial institution). If one company in the group is experiencing problems, other group companies will assist with finding a solution.

Since most companies in a chaebol are not publicly held, it is often impossible to determine the precise financial standing or structure of a particular company or of the group. Consolidated financial statements are not generally available and furthermore, Korean accounting practices are not always consistent with international practices. As a result, both Korean and foreign bankers and businessmen largely depend upon reputation in dealing with a Korean company.

In many cases, a foreign company will find that doing business with one of the chaebol (although not necessarily one of the super-chaebol) is the most desirable alternative. The chaebol have excellent relationships with the Korean government and financial institutions and have a large number of well-trained employees. Another advantage is that some chaebol are able to finance new projects internally. Chaebol are experienced in international transactions and have a network of overseas offices, are familiar with western management practices and have prior experience in dealing with foreigners. These large business groups are able to attract top quality employees because of their prestige.

There are also drawbacks to doing business with a chaebol. This is particularly true in the case of joint ventures. First, the chaebol may not devote its full energy and resources to the project since it may have other priorities. Second, the chaebol may indirectly use a profitable joint venture to support related group companies which are not profitable, by adjusting purchase prices paid in intra-group transactions. Chaebol inevitably purchase items from group companies even if the price is not competitive or quality is less desirable. You will not find a Daewoo elevator in a Lucky-Goldstar building and vice versa. Third, the chaebol may try to shift its less qualified employees to the joint venture. Whether a foreign company can control or manage these problems is largely dependent upon which party has equity and therefore management control over the joint venture, and how much the Korean company needs the technology and business experience of the foreigner.

In dealing with a chaebol, it is wise to determine early in the negotiations whether government restrictions prevent the Korean company from raising sufficient capital to fund the project. The Korean government has ordered chaebol to reduce cross-holdings in subsidiaries and affiliates, a policy implemented through the Fair Trade Commission. The government intends to reduce the size and influence of the chaebol. Under the Fair Trade Law, the total investment in all other companies on a book value basis by any company in designated chaebol cannot exceed 40% of the net assets of the company. Some exceptions to this rule exist so each case must be evaluated carefully. This rule can prevent a chaebol group company from making an investment in a joint venture unless it increases its net assets or reduces its investments. The same law prohibits cross-capital investment within chaebol companies and prohibits the establishment of certain holding companies. Other laws and regulations also restrict the ability of the largest chaebol to borrow funds from financial institutions.

Other major Korean business groups	
Group	Major business sectors
(1) Hanbo	Construction
(2) Kukdong	Petroleum products
(3) Hanyang	Construction
(4) Dongkuk Steel	Steel
(5) Miwon	Foods
(6) Korea Shipbuilding & Engineering	Shipbuilding
(7) Life	Construction
(8) Samwhan	Construction
(9) Shindongah	Insurance
(10) Halla	Construction
(11) Kohap	Textiles
(12) HS	Footwear
(13) Tongkuk	Textiles
(14) Pacific Chemical	Cosmetics and foods
(15) Taihan Electric	Wire and cable
(16) Samyang	Foods
(17) Poongsan Metal	Metal products
(18) Nhongshim	Foods
(19) Tongyang	Cement and foods
(20) Dong-A Pharma	Pharmaceuticals

Non-chaebol partners

Recently, a number of small- and medium-sized Korean companies have begun to emerge which have the capital and technical expertise necessary to compete with the chaebol. A major advantage of dealing with a smaller Korean company is that it is more likely to devote its capital and other resources to your business project since they have fewer alternative opportunities. Small- and medium-sized companies are also more likely to accept the foreign company's management control over the business since they are not as experienced or technically sophisticated. These companies may also benefit from special support provided by the Korean government to reduce the degree of concentration in the economy. Such special support usually takes the form of tax benefits and increased access to capital or loans with concessionary rates.

There are disadvantages to dealing with a Korean company which is not part of a chaebol. Smaller Korean companies may lack the real property and other tangible assets customarily required by bankers as collateral for loans. Unlike the west, well-documented profit opportunities (i.e., a business plan) may not be sufficient to raise capital for a venture. Smaller companies also have little experience with establishing accounting procedures and preparing financial records and projections. In addition, many small and medium sized companies are dominated by a key man. If this person were to die or retire, control may pass to someone who is not an acceptable partner. Another problem may be that the owner of the small company may need dividends for personal reasons when it is not in the best interest of the company. A foreigner must be careful when advancing capital to a small company since assets may be subject to prior security rights. In some cases, obtaining collateral security rights or an acceptable guarantee will be appropriate.

Whether one deals with a large or small Korean company, the most important single factor to consider when selecting a business partner in Korea is whether it has basic business objectives which are compatible with those of your company. This is particularly true in the case of joint venture projects. At the very least, the following objectives should be considered:

(i) short- and long-term profit (return on investment)
(ii) technology transfer
(iii) division of markets (particularly third country markets)
(iv) profit remittance or profit reinvestment.

Having consistent objectives as well as risk tolerance levels is conducive to a successful venture. The opposite is also true. For example, trouble often arises in a business venture involving a foreigner if the Korean company places more value on growth and market share than profits. The objectives of the business venture should be the first topics in any business negotiation.

Every Korean company has its own style and unique attributes. For example, the Hyundai group is known for its *go it alone* style of operation. Hyundai tends to hire expertise and build its products with its own technology. In contrast, the Daewoo group is known for working well with foreign companies in joint ventures, technology licences and technical cooperation agreements. Daewoo has joint ventures with GM and Leading Edge which have been very successful. Particularly when selecting a smaller company or a distributor, it is wise to choose a company with basic knowledge about the industry. Checking the company's influence with Korean government officials will be also useful.

Many of the most successful joint ventures and business collaborations in Korea occur when a foreign company with desirable technology and business skills approaches a chaebol or other Korean company which is not currently involved in the proposed line of business. Since the foreign investor is much more knowledgeable about the business in such a situation, the Korean company is more likely to allow the foreign company to make the decisions which will make the venture profitable. It will always be easier to train someone with no preconceptions about proper operating procedures than to re-train someone with their own ideas on the relevant procedure. Foreign companies which choose a business partner already manufacturing and marketing a similar product may find that the Korean company makes poor or misinformed business decisions due to their reluctance to change previous practices, technical specifications and marketing approaches.

The decision-making structure

The concentration of decision-making power in Korea is unparalleled in western industrialized countries. Rigid hierarchies exist in every Korean organization and significant decisions are made only by those at the very top level. That is not to say that the *working level* personnel can be ignored. They have an equally important role in reaching that decision.

Korean company structure

Board of Directors	Chairman
	President
	Vice-President
	Managing Director
	Director
Non-Directors	Department Chief
	Deputy Department Chief
	Section Chief
	Deputy Section Chief
	Branch Chief
	Staff

Some Korean companies have created sub-levels for some of the titles noted above (e.g., executive vice-president). Non-directors are commonly referred to in Korea as *working level* employees. Every position has a title, including ancillary staff such as drivers and messengers, which determines their status.

Koreans and the foreign press often describe Korean companies as being consensus oriented and eager to have all employees participate in decision-making. This is nonsense. Korean managers do not trust or listen to their subordinates to any greater extent than they may in foreign countries.

In fact, the most notable characteristic is their authoritarian paternalism. Managers expect obedience from employees who in turn expect paternal care from managers. All Korean organizations have a chairman or other supreme leader who has the power to make decisions quickly and decisively. Other employees of the Korean company will quickly work to implement the chairman's decision once it has been made. The decisions of the chairman and other top executives extend down to a level generally unheard of in western industrialized countries. Delegation of decision-making authority is done sparingly and to an extremely limited extent. What this means as a practical matter for the foreign businessman is that getting the business decision to the decision-maker is crucial.

The higher the level at which the foreigner starts in the Korean company, the faster the decision will reach the key person. For significant deals, it is best to at least start with someone at the director level and preferably the chairman. Typically, the objective is to obtain a broadly framed approval for the venture from the final decisionmaker with specific details to be determined by lower level employees who have direct working responsibility for the relevant business transaction. Once the broad mandate has been given, the working level Koreans should be contacted by more junior employees in the foreign company. The decision should then work its way up proper channels to the decision-maker It is useful to keep in mind that on the way up the decision-making chain, a foreigner will deal with many people who can say no to a deal but with no one who can say yes until he reaches the top level. If a foreigner has received a broadly framed yes at the top, it is unlikely that a no will arise at lower levels.

Finding the decision-maker in a Korean company is not a simple task. It should be remembered that an executive who is related to the controlling family may exercise more power than his nominal superior. Many Korean companies, particularly chaebol, have ex-government or ex-military officials in more senior positions than younger family members. Despite their more junior title, the family member may be more important in the decision-making process. Status can also be determined by closely examining certain cultural clues. When two Koreans meet, the relative depth of each person's bow is an excellent indication of relative status.

It is not an exaggeration to say that fewer than 1,500 people have the power to make significant business decisions in Korea. It should be noted that there is nothing sinister to most Koreans in this concentration of power. The power structure in Korea is entirely consistent with Confucian teachings and Korean values. While Koreans accept the stratified and highly vertical power structure, they will work as hard as they can to increase their position within the structure. It is this hustle to succeed and raise their relative status within the power structure and hierarchy that has made Korea grow and prosper so rapidly over the past 25 years.

Basic standards of decorum and business etiquette

While certain basic standards of decorum and business etiquette should be observed, making a mistake is not an absolute barrier to doing business in Korea. It is more important that a foreigner show himself to be a sincere and culturally sensitive person than establishing a high degree of knowledge of local culture. However, making even a token effort to understand Korean culture is important. Efforts to learn something about Korean history and the Korean language will always be appreciated.

The following is a list of basic cultural courtesies and business etiquette to keep in mind while doing business in Korea:

1. Avoid embarrassing a Korean in the presence of other Koreans. This is true even if the Korean is of low social status. There are always much more subtle ways of accomplishing an objective than causing a Korean to lose face in front of other people. If criticism becomes necessary, do it in private. In some cases, it may be helpful to use an intermediary to convey criticism, particularly with someone of high social status.

2. Do not undercut the authority of a more senior Korean by directing questions or responses at a more junior Korean. This may be frustrating at times since the junior person may more readily understand your questions or position on an issue due to greater English language proficiency or closer involvement in the deal. Many foreign businessmen make the mistake of directing questions and answers to the best English speaker. The senior member will lose considerable face in such a situation and the result will be counterproductive. One useful method of dealing with this problem is to schedule a working level meeting of the more junior members involved in the negotiation.

3. Most Koreans have three names: a family name (or surname) and a personal name (part of which denotes a man's generation in the family clan). Do not call a Korean by the first portion of his given name. For example, Moon Suk Chung would never be called "Moon" no matter how well you know him. After you have been introduced to him, you should call him Mr Chung. When you get to know him better you might call him "MS" or "Moon Suk." It is preferable to wait for an "invitation" before using a person's given name, except in the case of very westernized Koreans. If Mr Chung is a professional, you should refer to him by his title, for example, Lawyer Chung or Doctor Chung. A chairman, president or director of a Korean company should be referred to by placing his title before his family name. Similarly, high government officials should be referred to by using their title, for example, Director Kim.

A few additional facts about Korean names are worth noting. Koreans will place the last name first when referring to a person in the Korean language. Chung Moon Suk for example. However, in deference to western custom, the family name is often placed last when English is being spoken. Korean family names are easy to recognize since the vast majority of Koreans are named Park, Kim, Lee, Choi, Chung, Ahn, Cho, Chong, Han, Ku, Ko, Im, Oh, Noh, Shin, Yu or Yun, and because the given name has two syllables. Over 20% of the population has the family name Kim, Lee is used by 15% and Park, 8.5%. In total there are only about 300 family names in Korea.

It is seldom a good idea to suggest that a Korean call you by your given name unless he is very westernized. If you meet a Korean businessman's wife, remember that she does not take her husband's name upon marriage and instead retains her maiden name. She may be referred to as "the wife of [husband's family name]". Some very westernized Koreans adopt the practice of assigning the husband's family name to the wife to avoid confusion, but this is only for the westerner's benefit. Ironically, the custom of a wife's retention of her family name does not arise from any concept of women's liberation but instead arose historically because the woman was not considered worthy of assuming her husband's surname.

4. Foreigners in Korea will quickly adopt the custom of bowing in both social and business situations. For a foreigner, a bow from the waist to a 10 to 15 degree angle for around two seconds will be appropriate in most situations. The eyes should be lowered and hands kept close to the outer thigh. The angle and length of the bow should be adjusted depending upon the social status of the person you are greeting. The lower you bow and the longer you hold it, the more respect is shown. In general, foreigners should mirror the bow given by the Korean. After a short time in Korea, bowing becomes an almost unconscious habit. Korean men have in turn adopted the western practice of shaking hands. A Korean may use two hands in a handshake to show his particular pleasure in meeting you. In such cases, you should reciprocate with a two-handed return handshake. A foreigner should not offer a handshake to a Korean woman unless she is very westernized.

5. Seats at a conference table should be allocated on the basis of status. The middle seat should be occupied by the most senior member with decreasing authority as one moves to the sides of the table. The Koreans should be seated on the same side of the table, if possible, with officials from the foreign company seated on the opposite side according to rank.

6. If you have brought your wife on a business trip to Korea do not insist that entertainment include the wives of your Korean counterparts. It is polite to make an invitation, but impolite to insist that a Korean woman attend a function which would make her extremely uncomfortable.

7. Do not be surprised or insulted if you are invited to dinner or other social function only hours before it begins.

Koreans are not in the habit of issuing invitations well in advance.

8. Writing in red ink is reserved for situations where a problem exists. Never use red ink in signing a document or submitting any type of application. Writing a person's name in red ink is considered particularly improper since it has connotations of death for the person named.

9. Use the English language when meeting with a Korean business executive or government official. Do not bring a translator unless you are requested to do so. Koreans realize that English is the language of international business and diplomacy and take pride in their English language proficiency. Not using English will typically be considered to mean that you do not feel the Korean to be capable of speaking English. Certain foreigners who are thoroughly fluent in Korean and are quite familiar to the Korean to whom they are speaking, may use the Korean language. Some very senior businessmen and government officials will speak through a translator, but in such cases a translator will be provided.

10. Be on time for meetings with senior Korean businessmen and government officials, but try not to be more than a few minutes early. Generally, expect such meetings to end on schedule.

11. Respond as soon as possible to inquiries, proposals, correspondence and invitations. At the very least, immediately send an acknowledgment stating that you will send an answer as soon as possible. The pace of international business demands speed and Koreans will question your ability to perform if you do not show diligence. One of the most common complaints of Korean buyers is the length of time Western companies (particularly Americans) take to respond.

12. It will be considered polite to escort Koreans visiting your office to a conference room when they arrive and to offer them some form of refreshment. For Koreans, a person's office or desk is a place reserved for working and not for meeting guests. For this reason, foreign companies establishing an office in Korea should consider renting sufficient office space for at least one conference room.

13. Do not force your employees to take Saturday off from work. Korean businesses, banks and government offices are open until at least 1.00 pm on Saturday. Adopting the foreign practice of closing the office on Saturdays may cause considerable consternation by your Korean staff. More than one foreign businessman has insisted on non-working Saturdays only to find months later on a chance visit to the office that everyone but him has been there each Saturday as usual. Korean people place great emphasis on diligence and presence at the work place even though they may have little to do.

14. Be prepared to eat well while you are in Korea. Meals will stretch over a considerable period of time. If you are eating Korean food, it is generally advisable to let the Korean order for you unless you are very knowledgeable. Typically, the same dishes will be served to each person. Westerners will nearly always be served dishes which are unobjectionable.

15. Do not be overly anxious if invited to a *kisaeng* party while you are in Korea. Kisaeng are the Korean counterparts of the Japanese geisha. In ancient times, kisaeng were specially trained female courtesans who were trained in the art of pleasure since their childhood. Modern kisaeng are little more than personal waitresses who laugh, eat and drink with their customers. The proceedings at such a party are harmless and invariably good fun. Recently, some foreign women have been invited to kisaeng parties out of deference to western business practice and customs, but asking if your wife can come to a kisaeng party is generally not a good idea unless you need to convince her of the harmlessness of the proceedings.

16. Do not refuse an empty drinking glass which has been offered to you. You will be expected to take the glass and allow it to be filled with liquor. If you do not drink, politely explain to the Koreans present that you do not consume alcohol. Hold the glass in the air while it is being filled and try not to place it on the table until it is empty. It is a good idea not to take too long to empty the glass since the person who offered it to you will not drink until it is returned or he receives a glass from someone else. Do not pour liquor in your own glass. Try to fill the glasses of nearby Koreans which are empty or near empty. It will be considered good form if you in turn offer your empty glass to one of the Koreans. Be sure to accept and offer glasses with two hands rather than one. If you have had enough to drink, leave your glass full as an empty glass is a sign you desire more.

17. When entertaining in a bar or lounge, be sure to order a small appetizer or other food item with alcoholic drinks as it is not considered proper to drink alcohol without eating. Koreans view drinking alcohol as a social event and consider drinking alone to be improper.

18. Do not be bashful about participating in the singing which is inevitably a part of a kisaeng party, bar crawl or business dinner. Koreans will appreciate your good sportsmanship and will not expect a great vocal performance.

19. It is customary for Koreans to fight over the honour of paying the bill for food or entertainment. The concept of paying your own share is foreign to Korea and insisting that the bill be split is a major cultural blunder. While it may be difficult, you should make every effort to pay the bill when possible or to reciprocate at some later occasion. Hosting a meal on some future occasion is the best way to reciprocate. It will never be considered good form to ask how much food or entertainment cost. Rest assured that it was very expensive (especially if you have been at a kisaeng party or

room salon). Fortunately, Korean businessmen have generous expense accounts for entertaining.

20. If possible, inform a Korean of bad news in the afternoon rather than the morning. Koreans consider it impolite to do otherwise as it destroys the listener's *kibun*. This word is difficult to translate but generally means feelings or mood. To avoid damaging a Korean's kibun it is also a good idea to try to temper any bad news with some positive aspect. When a Korean is relaying unfortunate news to a foreigner he may temper or soften the news to a point where the message is totally lost to avoid damaging the listener's kibun.

21. Koreans have for centuries been called the *Ceremonious People from the East* by their Asian neighbors. When entering into an agreement with a Korean individual or company it is therefore appropriate to have a ceremony to mark the occasion. Formal signing of the documents should generally be followed by a small reception. If the signing is in Korea, the Korean party will usually arrange the reception. If possible, top management of the foreign company should attend the signing ceremony. Banners should be made, photographers present and mementoes distributed.

22. Do not press a Korean to talk a great deal during the meal. Koreans traditionally eat first and then talk later. The post-meal discussions may continue for many hours, though the lunch itself may only last an hour or so.

23. Individuals are judged more by their style of dress and physical appearance in Korea than in the west. The *dress for success* look is always appropriate for business meetings. Conservatively tailored grey and dark blue suits with subtle patterns should always be worn. Ties should be conventional, shirts white or blue and shoes black. At the very least, wear a jacket and tie for any business meeting.

24. While the western handshake has been adopted by the most Koreans, touching or back-slapping should be avoided (particularly if the Korean is female). Koreans generally need more personal space than westerners, so do not stand too close.

25. Non-verbal communication is very different in Korea than it is in the west. Avoid using body language which a Korean may not understand. The following are points to remember: (a) a shrug of the shoulders may not be understood; (b) a wink of the eye means nothing; (c) a weak or limp handshake by a Korean man does not convey timidity or weakness of character; (d) a smile may not indicate pleasure or joy and could mask entirely different emotions; and (e) the fact that a Korean may avoid eye-contact when speaking does not necessarily mean that the speaker is untruthful or unsure of what he is saying. In general, avoid pointing (especially with a finger extended) toward someone as this is considered impolite. A Korean doesn't beckon with a crooked finger for you to come toward him, but moves his hand toward his body with palm facing in and his hand down. This is also how a taxicab is hailed.

26. An invitation to a Korean home is a rare and interesting opportunity since Koreans generally choose to entertain in a restaurant. Always bring a gift for the wife of your host and if possible, for the children when you visit a Korean home for the first time. Remove your shoes immediately upon entering the home and put on the slippers provided. It is not uncommon for the wife to remain the kitchen and not join her guests for dinner. Do not press her to sit down. Her command of English may be very limited and she may be more comfortable not sitting at the dining table. Most homes are sparsely furnished by western standards but are very comfortable to visit.

The primacy of cash

There are three primary reasons why cash is king in Korea. First, Korean companies are geared or leveraged to an extent which far exceeds standards acceptable in western industrialized countries. The average debt to equity ratio of Korean companies listed on the Korean stock exchange is in excess of 4:1. It has been reported that companies in the ten largest chaebol have an average debt to equity ratio (excluding cross-equity investments) of 5:1. Debt to equity ratios of private Korean companies not associated with chaebol are much higher (particularly for very small companies).

Second, loan capital in Korea is very difficult to obtain for companies without collateral security. Because of the paucity and unreliability of financial information, Korean banks prefer to lend to large well-known borrowers. It may be difficult for a Korean company to obtain the necessary capital to enter into a venture unless the project is in an area favored by the Korean government. While Korean banks are ostensibly private institutions, they closely follow the policies and guidelines established by the government. For these reasons, capital is a relatively scarce commodity and not all companies have access to it at any given point in time. Third, real interest rates in Korea are very high and fees on loans are very large by foreign standards.

A number of consequences result from the relative scarcity of capital:

1. Projects which generate significant cash flows will be favoured. The cash is needed to meet debt servicing obligations on outstanding loans. This can often mean that sales growth and market share are more important than return on equity. It should be noted that since Korean companies do not have outside stockholders demanding dividends they can, in some cases, take a long-term view of a project. Whether a Korean company can take this long-term view is directly related to the cash position of the company.

2. The Korean government places a great deal of informal pressure on Korean companies not to declare dividends, since dividends paid to a foreign company and remitted

abroad reduce the government's foreign currency holdings. The Korean company will be reluctant to take cash profits out of a joint venture in the form of dividends and will prefer to reinvest profits in the business. Thus, if the foreign company has less than a 50% interest in a joint venture, dividends may not flow from the investment for a long time unless some type of contractual protection is provided. Growth is valued over profits by Korean companies because dividend payments are subject to taxation and because the shareholders may have no alternative need for cash. In any joint venture, dividend policy should always be decided in advance.

3. Significant cash payments will be required to engage in certain business activities. Most notably, residence and office leases typically require huge key-money payments (refundable deposits which theoretically produce a stream of income equal to some portion of the rent), or advance payment of the entire lease amount. Rents in Korea (and particularly in Seoul), are high in relation to other business expenses.

4. The western concept of paying for items using a personal or company cheque is almost completely unknown in Korea. The acceptable means of payment are cash, bank cheques and direct interbank transfers. Foreigners in Korea quickly become accustomed to carrying large amounts of cash in order to pay bills which would typically be paid by cheques in their own country. Another method of payment commonly utilized in Korea is the promissory note. The widespread use of this payment technique has caused commentators to describe Korean business as an island floating on a sea of promissory notes. These promissory notes generally comply with the requirements of the Korean Bills Act by being made on preprinted forms approved by the Korean Banking Industry Federation. Promissory notes executed in this manner are highly negotiable and must be protected as if they were cash. Failure to honor a promissory note is not only a serious breach of business ethics, but can subject the maker to criminal punishment. For this reason, a promissory note, particularly if notarized, is considered good security. Korean businessmen are careful to honour their obligations under promissory notes.

5. Korean exporters are likely to require a letter of credit before starting production. A letter of credit can be taken by the Korean company to a Korean bank and discounted immediately for cash which is then used to finance production costs. The government continues to encourage this form of export financing. Non-standard financing methods require advance government approval.

6. Korean companies understand the *time value of money* very well due to the high interest rates charged by banks and private lenders on the kerb market. Foreigners should always comply precisely with all formalities (particularly those imposed by the government) to avoid payment delays. The failure of a foreigner to comply with the smallest procedural detail is an opportunity to delay matters further and will nearly always be seized.

7. Payments on trade debts are typically stretched to the limit and settlements made 180 to 360 days after receipt of an invoice are common. Chaebol are notorious for withholding payments due to both domestic and foreign sub-contractors and for giving a promissory note as payment. Problems with a product or service should always be resolved before payment is made since bargaining leverage decreases dramatically (some would say to nothing) when cash is transferred. This is especially true if no subsequent dealings with the Korean company are planned.

8. Because Korean companies are so highly leveraged, Korean banks keep a close watch on the corporate activities of borrowers. This scrutiny is encouraged by the government as companies are not considered to be capable of managing their financial affairs. A prime bank is assigned to each company to assume control of it in the event of financial distress. Bankruptcies (liquidations) of substantial companies are very uncommon in Korea because the business is stabilized by the bank and then transferred to another company. Reorganizations similar to those which occur under *Chapter 11* of the US Bankruptcy Code also occur in Korea under court supervision.

9. Korean companies are comfortable with a much greater level of debt relative to equity than a foreign company. The Korean company believes that the level of debt creates mutual interdependence between the lenders and the company.

10. Because of the need to generate cash flow to meet debt servicing obligations, Korean companies devote their energies to increasing market share and revenues rather than raising their P/E ratio. This emphasis is encouraged by bankers who tend to make loans based upon market share and company size. Government officials also encourage growth over return on equity by awarding favours, privileges, benefits and subsidies to companies which meet export targets, even if the products are sold at a loss.

11. Financing sources for the activities of a foreign-owned enterprise or joint venture will vary depending upon the type of funds borrowed, the purpose of the loan and the foreign company's business relationships with Korean business groups. Since funds are always scarce, proper business connections are essential. Loans of won currency do not require government approval in most cases. However, the approval of lending officers of a bank is effectively tantamount to government approval. Loans in foreign currency require a government approval if the term is longer than one year and a licence if less than one year. Approvals will only be granted if the funds will be used to achieve targeted policy objectives. Typically, most foreign companies borrow funds from the branch of a foreign bank on an unsecured overdraft or promissory note basis.

12. The attention given to a project will dramatically

increase if cash payment is promised immediately upon performance. This fact can be used to great advantage in emergency situations. Situations where performance is declared to be impossible can often be resolved quickly by the promise of a cash payment upon completion.

Korean auditing and accounting practices

While considering the importance of cash to doing business in Korea, it is helpful to consider Korean auditing and accounting practices. The Korean Commercial Code requires that financial statements be prepared for shareholders. Companies listed on the Korean Stock Exchange must also file audited financial statements with the Korean Securities and Exchange Commission. Certain other large Korean companies must also be audited annually by an independent certified public accountant. While Korea's system of *Generally Accepted Accounting Principles* is based upon western concepts, significant differences exist. In addition, unique financial practices (such as the frequent use of promissory notes as a means of payment) mean that financial statements should be examined with care. Many small Korean companies do not maintain official books or are unable to present accurate financial information. One deals with these companies (and Korean companies in general) largely on the basis of reputation.

Korean negotiating style

Koreans have a reputation for being tough negotiators. They are aggressive, persistent and willing to exploit any weakness or opening. The success of any business venture in Korea is directly related to the outcome of negotiations with Korean businessmen and government officials. In order to negotiate successfully in Korea, a foreigner must be sensitive to values, behaviour patterns and psychology as well as to purely business matters. The following is a list of points which foreign businessmen should consider and tactics which they should expect to encounter during negotiations with Korean businessmen:

1. Proper preparation is essential in order to achieve an acceptable negotiated result. Expect the Korean company to know your company, your products and the industry and to ask penetrating questions. Have the relevant information at hand and be prepared for exhausting discussions of the business proposal, particularly if it involves a substantial capital investment. Inexperienced foreign businessmen are always surprised at the detailed questions asked by their Korean counterparts. Investigate the Korean company's business, markets, financial status and reputation prior to your first meeting. Research current economic conditions and government policies in Korea as thoroughly as possible.

The commercial section of your embassy and local foreign chambers of commerce in Seoul will be able to help you obtain this information. It may also be useful to approach a business consultant or other professional in this information-gathering process.

2. If the negotiators do not have decision-making authority (since significant decisions are made by the company chairman), you will be at a serious disadvantage because you are making binding concessions while they cannot. It is common for an agreement to be reached by the negotiators only to have the Koreans declare that everything is subject to approval by their superiors. This tactic is sometimes used to demand additional concessions. The best way to deal with this problem is to state in advance and at the conclusion of the negotiation that any changes made by superiors will open the entire agreement to renegotiation.

3. Some proposals may be introduced as suggestions by others (e.g., lawyer, law department or lower working level employees) before the negotiators have reached a final determination on their merit. In such cases, the proposals are *trial balloons* designed to discover your response. If you do not object, the proposals will quickly be adopted, and if you do, they can be withdrawn without a loss of face.

4. The Korean company will inevitably use a team approach for negotiations with additional support staff in the wings. Do not be overwhelmed by the size of their negotiating team, but bring as many people as you think you will need. It is generally prudent to bring individuals with expertise in each of the relevant areas under negotiation (i.e., manufacturing, sales, accounting and tax). In addition, staff at the home office should be ready to provide you with immediate support when necessary. Bringing lower level staff may help in moving the deal along since it may allow contacts directly between working level employees.

5. In most negotiating sessions, Koreans will speak both Korean and English. This can result in a considerable advantage for the Koreans since they can privately discuss a matter without leaving their seats (foreigners who speak a less well-known language will have the same advantage). At the very least, foreign negotiators should feel free to excuse themselves to discuss negotiation points in private.

6. It is not uncommon for Koreans to use brinkmanship to win a negotiating point. In other words, expect to receive ultimatums from the Korean company during the negotiation process. It is generally best to avoid giving ultimatums in return and to present arguments both for and against. In fact, many submissions to Korean government agencies (such as the tax office) must state both positive and negative arguments. Power plays or table pounding by a foreign company are seldom effective unless the company has something very desirable to offer the Korean economy. In responding to ultimatums made by a Korean company do not compromise your basic principles. By no means should the foreigner concede merely to get past an ultimatum proposal.

7. Do not feel compelled to fill in silences which may occur during negotiation. Businessmen from western countries are particularly uncomfortable with such silences and typically make statements to fill what they feel to be an awkward period, even to the extent of making unnecessary concessions. Being silent is particularly useful if the Koreans have made an unreasonable demand or proposal. The ensuing silence may cause the Koreans to feel compelled to compromise or offer a solution.

8. Do not let the Korean company know that you have a fixed date for leaving Korea or for finalizing the transaction. If the Korean company learns of your desire to complete the deal, it is very likely to use your anxiety as a lever to obtain more favourable terms. This technique of waiting until just before the foreigner leaves to demand a concession or provide a definite response is really just another form of brinkmanship.

9. The pace of the negotiations with a Korean company will inevitably lag behind the foreigner's expectations. The conversation at the beginning of a meeting is nearly always informal and often is not business related. Koreans believe it is more important to establish a relationship and will be focusing their efforts toward learning about the negotiators and their company before concentrating on the agreement.

10. Koreans are not comfortable working in English and will generally welcome your provision of a draft. When negotiating an agreement, a word processor is useful to get the document "on-line" so that a quick turnaround is possible. In general, agreements should be shorter than their western counterpart, easy to understand and broadly phrased. Time spent haggling over minute details is usually counterproductive.

11. It is vital to have proper legal advice in negotiations with Korean companies. The negotiated agreement must contain certain basic provisions if the foreign investor is to be adequately protected. A well-qualified and commercially oriented lawyer who can assist you in facilitating your business transaction should be utilized. It should be emphasized that Korean law is very different from foreign law. Furthermore, legal matters which are important in Korea may not be important in the foreign country.

Korean consultants may discourage the use of lawyers because most will be unfamiliar with commercially oriented attorneys. Although the number of lawyers admitted to practice in Korea is very small, every Korean company has a number of individuals involved in the negotiation process who studed law at university. A foreign company that does not receive equivalent advice will be at a significant disadvantage.

12. Be prepared for negotiating sessions to extend well into the evening. Morning sessions are less common and are invariably less productive than late afternoon and early evening negotiations.

13. It is unreasonable to expect business decisions to be made during an initial meeting. If you must have an immediate response, always send a written proposal well before the meeting.

14. Koreans are sensitive to the ambience of the negotiation. They will react unfavourably to a person they consider to be impolite or harsh and appreciate someone who is sincere, polite and persistent in their approach.

15. Negotiations will always continue after the agreement is executed. In fact, the really important negotiations take place as the venture progresses. The real substance of the relationship arises as the two sides negotiate concerning events which arise on a day-to-day basis. The existence of this need to constantly negotiate matters means that a foreign manager resident in Korea is particularly crucial to the success of the business venture.

16. If negotiations need to be terminated, great care should be taken since a poorly executed termination can make future business impossible or at least very difficult. Try to avoid fixing blame on the Korean company and avoid characterizing the proceedings as a failure. Always keep the reasons for any breakdown of negotiations out of the public domain to prevent a loss of face, and strive to characterize the termination as a temporary cessation which may be resumed when economic conditions change.

The fluidity of contractual agreements

One of the most difficult problems when dealing with Korean businessmen is their markedly different attitude toward the sanctity of a written contract. The Korean party generally views the contract as a ceremonial memorial to the terms of the contract on the day of signing. Korean businessmen will expect problems to be addressed as they arise on a case-by-case basis and the nature of the agreement to change as economic conditions vary. In other words, a relationship rather than a fixed set of terms. In contrast, the foreign businessman views the contract as defining the terms of the deal and expects to plan for the long-term on the basis of the contract. The foreigner tries to anticipate problems before they arise and expects problems to be resolved pursuant to the contract. Thus, it is unsurprising that a conflict arises when the Korean party approaches the western businessman and expects to have contractual terms revised to reflect a changed economic condition. The Korean will expect that the relationship between the parties will allow a mutually acceptable solution to be achieved.

Unfortunately, a foreign businessman who approaches his Korean counterpart for a contractual change to reflect developments which are unfavourable to his company is often met with silence or hostility. At times, a legal double standard appears to exist under which Korean companies expect the Korean view of a contract to prevail when they desire a

Korean companies provide a favourable environment for their workforce, pictured here is the ornamental Carp pond at the Hyundai shipyard at Ulsan

change, and the foreign view to prevail when a foreign company desires a change. While there is no easy solution to this problem, creative negotiating can often produce a solution which benefits both parties. For example, the foreign company may be able to obtain a faster settlement of accounts or a longer contract term in return for a price concession requested by the Korean company. In any event, the foreigner must make it clear that he will not accept this double standard view of a contract.

One other reason why Korean companies and businessmen are so baffled by a Westerner's obsession with contracts is their more feudal view of the world. In a vertically oriented society composed of a complex network of clans, businesses and social groups, there is no need for a contract to create bonds between the parties. In Korea, the inferior party (for example a domestic servant) is expected to diligently perform services for his superior in return for paternalistic benevolence. An example of this type of relationship is the bond which exists between small Korean sub-contractors and the large chaebol which produces the finished product. The sub-contractor is at the mercy of the chaebol, but can rely on a degree of paternalistic care from the chaebol. A contract is only necessary between Koreans where there is no relationship between the parties or where both parties are equal. As stated previously, neither of the two situations exist when Koreans are involved, since some characteristic always makes one party superior to the other.

Notwithstanding the apparent disregard which Korean businessmen have for a written contract, it is still extremely important to carefully negotiate the agreement for two reasons. First, the contract presents a binding agreement in Korean courts and should provide a framework for terminating or winding-up the venture in the event of an unresolvable deadlock. Second, a well-drafted contract serves to define the business relationship and ensures that both parties fully understand the business transaction.

Special considerations are necessary when negotiating and drafting a contract with a Korean company. The western concept of slipping clever contract terms past the other party and later turning them into victory simply does not exist in Korea as a practical matter. Full and complete disclosure of all details of the deal will always result in fewer disputes and a better business result. Agreements should be written in clear and simple language. Koreans view a contract as relating to events due to take place in an uncertain future. They will want the contract to be as general as possible in order to provide flexibility to deal within this uncertainty.

Resolving disputes privately

Korean businessmen believe that the relationship between the parties concerned should by its nature prevent any conflict or dispute from arising between parties. If parties are forced to resort to formal legal process (i.e., the courts or arbitration) to resolve the dispute, a public admission has been made that the relationship has failed. As a result, if a dispute becomes public knowledge, both parties will be considered at fault regardless of the merit of any one party's position. This aversion to litigation and the high value placed on the maintenance of harmony in social relations arises from a number of sources. Most important are the Confucian principles which teach that inferiors must be loyal and superiors benevolent. Other factors contributing to the lack of litigation include a low level of rights consciousness and the tradition of delegating dispute resolution to non-official mechanisms.

When a dispute occurs, often the first instinct of western businessmen (particularly Americans) is to consider litigation. For a number of reasons, the use of litigation in Korea is almost never advisable. Most importantly, litigation (and to a slightly lesser extent, arbitration) is both time-consuming and expensive. Furthermore, it is always difficult to obtain complete justice from a foreign tribunal. There is nothing sinister in this statement. Korean judges are not trained in foreign law (which is often the governing law under the contract), do not always understand English (or other foreign languages), tend not to be commercially oriented, and have the natural (i.e., human) tendency to favour their own citizens. In practice, a Korean court's judgment often results in splitting the difference between the parties' respective positions. Litigation intended to force the Korean to pay damages is entirely inconsistent with a continuing business relationship. Therefore it is generally advisable to consider all possible alternative methods of resolving a dispute.

One such method involves bringing in executives from both companies who have no prior involvement in the relevant problem. Often the fresh perspective of these executives serves to facilitate a mutually acceptable solution. The newly involved executives have no entrenched position to defend and can concentrate on finding a resolution to the dispute which benefits both parties. This technique can be particularly effective in Korea since dispute resolution can be blocked by a person's desire to avoid losing face despite the economics of the situation.

Conciliation also provides an alternative to litigation for many types of disputes. Conciliation procedures can be purely private or imposed upon the parties by courts or arbitrators. As a practical matter, as noted above, conciliation is the most important method of dispute resolution in Korea because it is the only alternative which is likely to allow the parties involved to continue doing business. Furthermore, conciliation is a method which allows disputes to be resolved quickly.

Dispute conciliation in Korea which involves a foreign party is particularly difficult because the foreigner does not have an established place in the social hierarchy. Thus, in nearly all cases the most effective method of resolving the dispute is to adopt the practice utilized by Koreans when two parties to the dispute are of relatively equal social status. In such a situation, a mutual friend or business associate generally serves as the conciliator of the dispute. Under this method, a third party who has equal or greater social status than the parties should typically be called upon to conciliate.

Choosing an appropriate conciliator is crucial to achieving a mutually acceptable solution to the dispute. Obviously, the conciliator must be someone who has no business or other tie to either of the parties to the dispute. For business reasons, it is generally preferable to select an individual from outside the relevant industry or at least someone who is not involved in the relevant line of business.

Conciliation is not mandatory. Either party may refuse to participate in conciliation. As a result, there is no set procedure and the parties are free to establish any structure they desire. This freedom allows the conciliation process to meet the business needs of the parties. It is extremely important that the conciliation remain a private matter. Once a Korean businessman has adopted a position which is known to the public, it will be difficult to convince him to back down to reach a settlement. The conciliation procedure should be simple and a fixed time limit established. Both parties should submit written descriptions of their position together with copies of all relevant documents without applying rules of evidence or procedure. Copies of the written submission to the conciliator should also be given to the other party. In theory, neither party has an incentive to present an exaggerated or false description of the facts, since the conciliators have no power to bind the parties to any particular resolution to a dispute.

In practice, the result of the conciliation will almost never produce a clear-cut winner and loser of the dispute. Instead, the outcome is likely to be somewhere between the positions of the parties involved. It will often be a less than perfect solution but it may be the only solution which allows the parties to continue doing business.

The benefits of mediation and conciliation should not be confused with arbitration. Mediation and conciliation are non-binding methods of persuading the parties to reach a settlement of the dispute. While Korea has an arbitration system roughly similar to that which exists in western countries, there is relatively little experience in Korea concerning arbitration. Foreign companies which agree to arbitrate disputes with the Korea Commercial Arbitration Board (KCAB) will face some difficulties in obtaining an entirely objective decision. KCAB arbitrators must be resident in Korea, the compensation is minimal, and foreign arbitrators who are residents of Korea depend upon government favours in connection with their own business. Arbitrating a dispute and enforcing an award can take from three to five years. Any arbitration clause providing for arbitration in Korea, should among other things, explicitly provide a method for choosing qualified arbitrators. If possible, a neutral forum for any arbitration proceeding should be specified.

Communications

Effective communication can prevent many problems from getting out of control and even if disputes arise, there is no substitute for finding a solution. Communication is particularly important when doing business in Korea since a Korean will always be reluctant to relay bad news to a friend, customer or business partner. Koreans possess an inherently over-optimistic outlook which can radically contradict reality and harm a business relationship in the long run.

One of the most significant barriers to communication is the Korean language itself. Although this is a significant problem for western foreigners, it is also a problem for those who speak Chinese or Japanese. While Chinese characters are a part of Korean written script (together with a uniquely Korean writing system), and while some elements of syntax are shared with Japanese, the Korean language is unique. Obviously, major western languages, such as English are radically different. In Korean, the verb comes at the end of a sentence, plurals are seldom used and word choice will vary depending upon the social status of those involved. The Korean language is much more subtle and indirect than western languages, creating further confusion.

Although Korean businessmen will have studied English for many years, the foreign businessman should not assume that his Korean listeners have easily understood what was said. Koreans rarely give non-verbal clues to indicate their comprehension and will be reluctant to admit that they have not understood. Repeating and summarizing your key points during a conference will facilitate mutual communication. It is also good practice to put into writing any numbers (especially to clarify millions and billions), instructions, matters discussed, points agreed, and so forth.

Koreans prefer not to contradict anyone and will tend to avoid taking a definitive positive or negative position in response to an inquiry. As a result, many foreign businessmen incorrectly interpret signals given or statements made by Korean businessmen. This characteristic has caused many foreign businessmen to leave Korea thinking an agreement has been reached, when the actual situation is just the opposite.

Investigate production capability and monitor quality standards carefully

Korea's technological capacity has been hindered by a lack of trained engineers and the brain drain. Despite efforts to solve this problem, Korea continues to have a shortage of qualified engineers. Unfortunately, many of Korea's best and brightest pursue career paths which move them toward a position in the government and business bureaucracy. Of the talented science and engineering students who study overseas, a large percentage fail to return to Korea and instead take jobs in advanced countries like the US.

In considering their technological capacity, it is wise to remember that Korean businesses have been known to bid on a job or quote a price which obtains the contract and only afterwards begin to consider whether the job can be done at the agreed price. If the bid is later determined by the Korean company to be too low, quality will inevitably be sacrificed to produce an acceptable profit margin. For this reason it is wise to suspect and carefully examine bids which are unreasonably low. In some cases, the Korean company will later demand changes in the contract or even refuse to perform without additional compensation. A fine line exists between giving due credit to the resourcefulness of the Korean company and rejecting the Korean company's proposal as being unrealistic. There is no substitute for actual experience in the Korean market in drawing this line. If the foreign company does not have actual experience in Korea, it should seek out someone who can provide the necessary skills and experience.

The time to monitor quality is while production lines are being established and during production. Koreans are adept at fashioning techniques to reduce costs by taking short cuts or using lower-grade materials and in devising a quick-fix to problems. While effective in the short run, such techniques can cause a long-term disaster. It is nearly always a worthwhile investment to have foreign engineers from the foreign company present during production or equipment installation.

An inspection certificate should nearly always be required before a purchaser is allowed to negotiate a letter of credit. Once a sub-standard or short shipment has been made and purchase money released to the seller (the typical case in a letter of credit transaction), it is not a simple process to obtain appropriate compensation. The cost of increased quality inspection should be considered by the foreign businessman when deciding whether to base production in Korea or for that matter anywhere in Asia. In addition, a safety margin should be built into cost estimates to reflect potential quality problems or production or delivery delays. Small cost advantages have a way of quietly disappearing.

Korea still has a major shortage of engineering talent and is reliant upon technology licensed from foreign countries. Korean companies are very advanced in the technology used in consumer electronics, but have simply not had enough resources to devote to research and development. The large chaebol have devoted resources to research and development but have lagged far behind western standards. Much of the research which has been performed in Korea has related to the assimilation of foreign technology rather than the creation of new technology.

Korean tax laws are unique and in some cases unfamiliar to foreigners. For example, the value added tax imposes strict and complex requirements on taxpayers. In addition, without proper registrations or approvals, a foreigner may find it impossible to obtain the proper residence visa or even install a telex or telephone or purchase an automobile. The requirements of the Korean government, while formidable, are fairly easily overcome with proper assistance. However, the foreign businessman should budget for the time and money necessary to comply with these requirements.

Employee selection

Foreign managers and technicians

It is nearly always prudent to have one or more employees of the foreign company resident in Korea. By being in-country and on-site, the employee can spot problems as they arise and keep abreast of any developments.

A foreign employee can keep the head office fully informed of events in Korea, assisted by other foreign businessmen who form a small but active business community in Korea. The foreign employee should participate in local organizations of foreign businessmen (such as the foreign chambers of commerce) to better take advantage of this source of information. Joining one or two social or athletic clubs (such as the Seoul Club) will also be useful for the employee both personally and professionally.

Local employees of the foreign company will also be responsible for nurturing relationships with the Korean partner or supplier and developing a network of local contacts and sources of information. In addition, particularly if he is involved in selling products to Korean companies, he will be expected to participate very frequently in male-only, after-work social activities until late at night. While a foreigner need not participate every night in such sessions, regular attendance is essential to success in business in all but rare cases. This will require the employee to sacrifice large amounts of time (particularly in the evening) which he would otherwise spend with his family.

Experience in Korea, other Asian countries (or anywhere overseas) will be an advantage, but not essential. From an operational standpoint, the individual will need to be a person who can make decisions without head office instructions and who can get things done. He will be involved in a wide range of activities (e.g., sales, marketing, finance, legal, accounting, research, advertising, and personnel), and thus a generalist is preferable to a specialist. The most effective foreign managers are those that push for the best result possible given the tools available and the human resources involved. Those managers who try to enforce exact standards in non-essential areas generate only frustration and tension. The foreign employee should also be flexible in his attitude toward management techniques, since foreign approaches do not always work in Korea.

A question which arises in selecting a foreign manager is whether he should speak Korean. If the person speaks fluent Korean this would be an advantage. Even less than fluent Korean language skills can help a person adapt to Korea as well as gain acceptance. It should be remembered that the language of business in Korea is English. Translators are seldom used when conducting business except when dealing with very low level employees. Knowledge of the Korean language is only one factor to be considered in the selection process.

Since age commands some degree of respect, the foreign manager should not be too young. As a rule, the manager should be at least 30 before being sent to Korea in a senior position, but since Korean businessmen are younger than their Japanese counterparts it is not necessary to send a man in his 50s or 60s.

It will take a foreign employee a considerable amount of time to establish the human relationships necessary to do business in Korea. Too many foreign companies transfer an employee just as he is becoming effective. It takes about a year to establish a basic business relationship. In addition, when rotating staff, a substantial overlap of the incoming and outgoing employees should occur so that accumulated knowledge, experience and contacts can be passed on. An overlap period of from three to four months is generally appropriate.

Past experience has shown that a foreign manager will be more effective if he has had previous contact with the Korean company or companies with which you are doing business. If the foreign manager is being sent abroad to head a new joint venture it is important that he be involved in the actual negotiations which result in the venture. This will strengthen the contract itself since promises are considered to be made between persons involved as well as between the relevant companies. In addition, involvement in the negotiation process acquaints the foreign manager with the Korean company's style and objectives.

Korean employees

Recruiting top-level managerial employees has many difficulties. Rather than citing reasons, it is more useful to identify what attracts a Korean to a foreign company as well as what discourages him from accepting such a position. Among the attractions of a foreign company are higher wages, shorter working hours, fewer office obligations during unpaid hours and advancement based upon merit. The disadvantages of working for a foreign company include little or no chance of obtaining a top position, lower status (in most cases), increased risk of being fired (perceived) and increased communication problems.

Particularly if the foreign company is an importer or selling goods on the domestic market, training Korean employees will be crucial to its success in Korea. An effective foreign manager will make a major effort to make the Korean employees feel part of the company family or team. It is often useful to send key Korean staff members to the foreign country for training to build staff loyalty. A promise to send a staff member to the foreign country for training is a powerful motivator, since obtaining permission to travel overseas is very difficult to obtain without sufficient justification.

Properly structuring a foreign company's Korean work force requires sensitivity to the expectations of the employees concerned. While foreign companies need not adopt a Korean organizational and promotional structure to be successful, it is necessary to consider employee expectations. For a Korean, the most crucial factors in assessing *promotability* within a business organization are educational status and length of service.

As Korean social structure is a hierarchy where individuals are born into a certain class based on family and other affiliations, upward movement is possible only through education. Thus, education is crucial in determining an individual's personal and economic prospects. It is common for families to move their residence to secure a place in a prominent Korean school, most notably a high school where academic pressures are particularly intense.

Most school and free time is spent studying for the extremely rigorous and competitive university entrance examinations which emphasize rote memory. Success in this university entrance examination is of paramount importance. A high score and admission to a top university can almost guarantee a student a lifetime of success (or at least a lifetime without failure). Once the student is in the university, the rigorousness of the educational experience declines markedly, and he or she can coast to graduation without significant effort.

Since the university attended by a Korean employee is one of the most important factors in determining his value as an employee, it is useful for a foreigner to understand the relative position of such schools. While opinions vary, consensus opinion ranks Korean universities as follows: (1) Seoul National University; (2) Yonsei University and (3) Korea University. Other top universities (not listed in any particular order) include Sogang, Hanyang, Sungkyunkwan, Kyonghee, Dongkuk, Chungang, Kunkuk, Dankuk, Pusan National and Hankuk. Ewha is the finest woman's university. Many Ewha graduates act as staff for foreign companies. A large number of influential Koreans are graduates of the Military Academy and the Military Staff College. The high school attended by a Korean is also very important. Top high schools among the current generation of Korea's leaders include Kyonggi, Seoul, Kyongpook and Kungnam. Care taken in selecting the right Korean employees will pay large dividends over time.

Total production of computers doubled in 1987 to US$1.5 billion

Because Koreans have worked so diligently to high school level, basic mathematics and science skills of Korean workers are excellent. However, the type of management and problem solving skills typically developed at a university level in the US or other western countries may be lacking. For this reason, Korean companies have their own training programmes. Workers destined for senior management are typically sent to foreign graduate schools for further training.

The second factor in determining the expectations of Korean workers concerning salary and advancement is length of service. In general, because of the importance placed on age in Korean society, those with more seniority are expected to be advanced to senior positions. Upsetting the expectations of Korean employees by promoting a younger or less educated person can create a lack of harmony and therefore serious problems within the organization.

Korean employees as a rule do not expect to be consulted about most decisions concerning the company. Orders and instructions from superiors are accepted without question in nearly all cases. Management, however, is expected to exercise paternalistic care toward employees in making decisions. In particular, employees expect management to make decisions in a manner which does not adversely affect harmony within the organization.

Foreign companies face unique problems in motivating Korean employees. While Korean workers are diligent and loyal, social constraints can make it difficult for a foreign manager to motivate and control employees. Simply put, western motivational techniques may not be appropriate for Korea. Often, a true bonus wage system based upon merit produces the greatest productivity from Korean workers and managers.

One major factor working in the foreign company's favour is the large pool of well-educated but relatively under-utilized female workers. Korean women employees (who in many cases are graduates from prestigious universities) can be great assets in staff positions. However, Korean women have had some difficulty operating at a management level because of the need to interact with Korean men at other companies. One disadvantage of hiring women employees is that turnover is much higher, particularly for girls who are not yet married.

Technology transfer

Koreans have traditionally believed that the reward for innovation should be cultural esteem rather than financial reward. This attitude springs from the Confucian belief that the pursuit of profit is not a proper role of a refined gentleman. This reluctance to recognize the financial value of technology is reflected in the Korean government's attitude toward technology transfer from foreign companies. First, the Korean government will not allow a foreign company to receive equity in a joint venture in return for providing technology. Only cash or in-kind contributions to the joint venture are acceptable in return for equity. This can create significant problems for technology-rich, but cash-poor foreign companies desiring to do business in Korea. The usual approach used to surmount this problem is to have the foreign company make a cash contribution in return for shares and then receive a licence fee for the technology (often in the form of a disclosure fee plus a running royalty). Second, the government strictly limits the amount which can be paid by Korean companies in return for technology and the duration of technology licences. Foreign companies with advanced technology must consider whether the restrictions placed on royalties and licence terms by the Korean government justify a technology licence. Often, the royalty rates are so low that only less than state-of-the art technology is suitable for licensing. If other Korean companies have already licensed similar technology, government approval of the licence may not be granted on the grounds that such technology is domestically available. This should give foreign companies a clue as to whether the technology will be shared by the licensee with other Korean companies, despite any contractual restrictions to the contrary.

Another problem with technology transfer in Korea is maintaining control of the technology once it has been transferred to the Korean company. Korean patent, trademark, copyright and other intellectual property laws have only recently been strengthened in response to foreign pressure. Despite these recent amendments, significant gaps in statutory protection exists. Most importantly, certain types of product patents are not available. An even greater problem exists in relation to unpatented technology. Much of today's most advanced technology takes the form of trade secrets and know-how. The Korean government will rarely allow restrictions on the use of such non-patented technology once a licence agreement expires. As a result, control over the technology will be lost by the foreign company upon the expiration of the licence. Furthermore, as a practical matter, even if contractual or statutory (i.e., patent or copyright) restrictions exist which prohibit the use of technology, it is not possible to obtain an enforceable injunction preventing use of the technology since Korean courts do not have contempt powers.

Control over technology transfer is also crucial in maintaining both the continued viability and management control of the joint venture. Care should be taken to avoid the *conduit-company* syndrome when a foreign technology-rich company joins with a Korean company to form a joint venture. The foreigner demands and obtains equity and management control in return for the technology, but is left with a *Trojan horse* corporate shell after the technology has gradually been transferred to the Korean company through the conduit. Often the Korean company will demand equity and managerial control of a joint venture once the basic technology and management skills have been mastered. The only sure way to avoid this problem is to continue to update and improve technology and management skills so incentives remain to continue the joint venture.

It is instructional to examine the view of Japanese companies concerning licensing technology to Korea. In general, Japanese companies have only been willing to license out-of-date technology or technology in sectors deemed to be sunset in nature to Koreans. The Korean press is full of complaints about the unwillingness of the Japanese to license technology. Japan's companies have, as a general rule, licensed technology only when they can use the transfer as a method of selling goods or components associated with the technology to Korea. Japan has seen first hand how quickly a competitor can be created in another country through a technology licence.

Major policies and objectives

Exports

Exports from Korea will continue to be encouraged by both business and government. The existence of a foreign company exporting from Korea will, as a result, be substantially less difficult than that of an importer or participant in the domestic market. Procedures for exporting goods are simple and government assistance is easy to obtain. Tax benefits, loan guarantees, subsidies, grants, duty drawbacks and other advantages will be granted to local companies (including those wholly or partially owned by foreigners) which export products.

Imports

Imports into Korea will generally be permitted for goods which increase production and facilitate exports. Importation of non-essential goods (i.e., luxury goods or consumer goods) will continue to be restricted. A few non-essential goods will be imported as a concession to countries with which Korea has a large trade imbalance, but their amount will be insignificant. Like Japan, numerous non-tariff barriers will continue to limit imported goods. These barriers will include licenses, approval requirements, government standards, quotas and informal (but extremely effective) government guidance and pressure.

Foreign exchange

Foreign exchange will be strictly rationed. Korea's Foreign Exchange Control Law is one of the most restrictive and best enforced in the world. Penalties for violating the Foreign Exchange Control Law are extremely harsh and can even include the death penalty. Activities which increase foreign exchange (i.e., exports) or save foreign exchange (e.g., import substitution) will be encouraged and subsidized. Activities which result in the expenditure of foreign exchange will be actively discouraged. Ownership and trading of precious metals and stones will also be severely restricted since such items can be treated as an equivalent to cash.

Technology transfer

Projects which result in the transfer of technology to Korea or the development of local research and development capabilities will be encouraged. Foreign businesses which are willing to transfer technology to Korea and train Korean workers will be granted significant benefits and preferences.

Foreign domination

Fear of foreign domination (often translated into ultra-nationalism) will restrict the ability of foreigners to engage in certain activities. For example, the ability of foreigners to control management of the joint venture and to own land is severely restricted by the Korean government. Limits are also placed on the ability of foreigners to purchase shares in public markets to prevent a loss of control.

Trade policies

Korea is liberalizing its trade policies (albeit slowly) as most industrialized countries move in the opposite direction. Foreign companies must take care not to be caught in the machinations of governments as they implement their trade policies. For example, a foreign businessman must be careful not to be caught investing in a protected domestic industry when the door is opened to more efficient foreign competition. Determinations must be made on a case-by-case basis, but in general terms the process of liberalization will be political.

Attitudes to the west

Korea is one of the few Asian countries which has no history of being colonized by western powers. As a result, a significant amount of goodwill exists for businessmen from western countries. This effect is magnified by the trade surplus which Korea has with such countries. This fact can be used to great advantage by western countries in competing for business against Japanese companies. This is particularly true for Americans because of their role in the Korean War and because their country assumes a large share of the Korean defence burden. The contribution of the US military in Korea has been estimated to provide as much as US$2 billion in benefits to Korea.

Natural resources

Projects which increase Korea's access to natural resources will be encouraged by the Korean government. Much of

The opening ceremony of the 1986 Asian games, hosted by Korea, was a forerunner to the 1988 Olympics

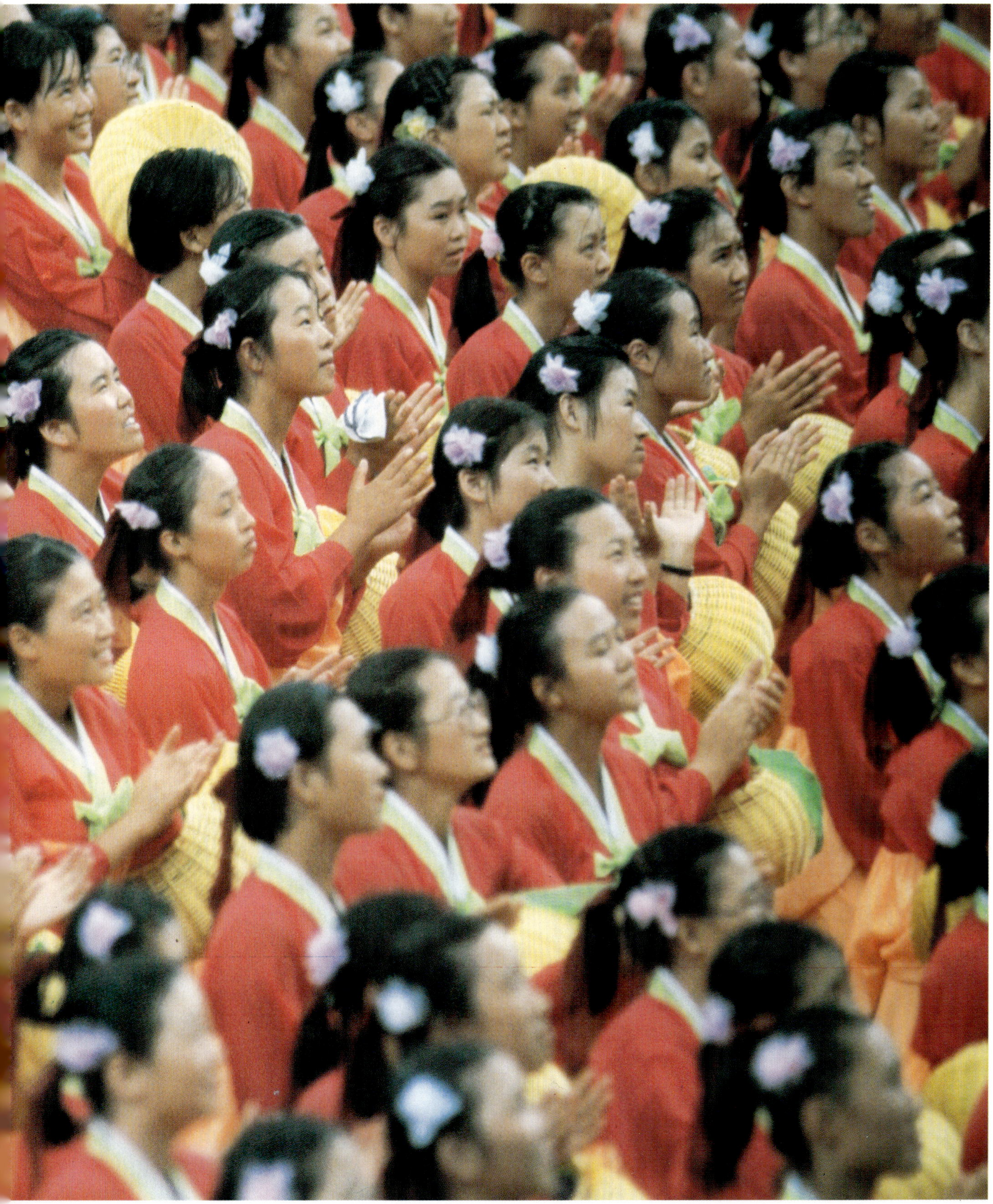

Korean investment overseas will be concentrated in this area to produce stable sources of supply. Joint ventures between Korean and foreign firms will be pursued in resource-rich nations.

Emigration

The emigration of Korean citizens to foreign countries will be encouraged to reduce population pressure and to generate foreign exchange from overseas remittances.

International presence

For defence reasons, Korea will strive to create investments in its economy by as many countries as possible to reduce the chance of a north Korean invasion, the theory being that a large international presence will deter aggression from the north.

Localization

Because of Korea's huge trade deficit with Japan, projects which reduce Korea's dependence upon Japan will be favoured by both business and government. Efforts will be made to produce Japanese-sourced parts and components locally. This localization process should accelerate as each year passes. However, because of past business practices and a lack of small businesses, the dependence on Japan as a supplier will not be easy to break. For example, in terms of value, 40% of Korean consumer electronic products are composed of Japanese parts and components.

Defence

Korea will continue to devote a considerable portion of its resources to defence (in excess of 6% of GNP). A significant effort will be made to localize manufacture of defence-related products to secure a stable source of supply, decrease outside influence and to increase security.

Rationalization

The Korean government will continue to target specific industries for support or rationalization. The government will look for industries which have forward and backward linkages. Such targeting may or may not work and non-targeted industries could succeed despite the lack of support.

Foreign debt

The Korean government will strive to reduce its foreign debt and to create a substantial capital surplus. As a rule, the Korean people like to buy items for cash. Being in debt is not favoured by Koreans either on a national or personal level. Given the recent current account surplus experienced by Korea, the nation may be a net creditor in the early 1990s.

Inflation

The Korean government will keep a tight rein on inflation, sacrificing growth to some extent if necessary. Because of previous double-figure inflation, Korea's monetary policy will be more like that of West Germany (which also has experienced post-war inflation) rather than the US.

Going public

Companies will be encouraged and coerced into going public (issuing shares to the public) to create a wider stake in the economy for the middle class and to promote domestic political stability. Korean companies will not change their habits easily and thus there will be a general shortage of share script relative to demand. This will mean investors will pay a premium for shares. As incomes continue to increase above subsistence levels, a shortage of investment opportunities will buoy the stock market. Companies will also be encouraged to hire professional management and to reduce nepotism.

Diversification

Efforts will be made to diversfy both import sources and export markets to protect the economy from changes which occur outside Korea. Particular emphasis will be placed upon reducing dependence upon Japan as a source of intermediate materials and capital goods.

Business growth

The growth of so-called small and medium industries (which are really fairly large businesses) and true small businesses will be favoured by the government. Their prosperity will increase technical mobility, diversify the economy and increase the distribution of wealth. The government will restrict the ability of large Korean business groups to impose harsh terms on small sub-contractors.

Government's role

The government's role in business will be gradually diminished (in relative terms), particularly in government-owned corporations, to promote efficiency. The operative word in this process is gradual since the Korean government bureaucracy is well entrenched and reluctant to part with power. However, the rate of growth and increasing sophistication of the Korean economy will soon overwhelm even the very competent and zealous Korean bureaucracy.

Korea's unique qualities

While experience in other Asian countries will be useful for doing business in Korea, it does not equate with experience in Korea. The culture and temperament of the Korean people are very difficult from other Asians. Furthermore, the business climate in Korea is different from other Asian countries in many respects. The foreign businessman should take care to avoid making assumptions based upon what he has learned elsewhere in Asia, particularly Japan. Assuming that what one knows about Japan can be applied on a wholesale basis to Korea is a fallacy and can be extremely counterproductive. There are very significant differences between the economies, trade policies, business practices, cultures and people of the two countries.

A certain element of mistrust for Japan permeates Korean society as a result of Japan's past repression of the Korean culture and the Korean people. It is precisely because of this mistrust of the Japanese by Koreans that a treaty normalizing relations between the two countries was not signed until 1965, 20 years after the end of the Second World War. The foreign businessman would do well to remember that informing a Korean that some aspect of his business, government or lifestyle is like the Japanese is at least embarrassing and usually insulting. The Korean people take great pride in the fact that much of Japanese culture can be traced to Korean origins (or at least travelled through Korea on its way from China).

Major differences between Korea and Japan

1. Korea has a much smaller domestic population (and therefore a smaller domestic market) than Japan. The population of Korea is approximately 41 million while the population of Japan is approximately 120 million. In addition, Japan's GNP is sixteen times greater than Korea. Because of this smaller domestic market, exports are much more crucial for Korean businesses in terms of utilizing economies of scale and recovering research and development costs. Korea is much more dependent upon markets in industrialized countries (particularly the US which receives over 40% of total exports) than Japan was at a comparable stage of development. Japan's exports in the 1950s were about evenly split between developed and underdeveloped countries. Korea is more vulnerable to increased competition from the People's Republic of China. It is illustrative to note that the marginal increase in the growth of the People's Republic of China's GNP in 1984 was greater than Korea's total GNP that same year.

2. Japan's development into a leading industrialized country occurred when foreign competition in western markets was considered largely to be a nuisance by domestic industries. In contrast, companies in western industrialized countries now consider foreign competition to be life-threatening. Korea is developing at a time when other newly industrializing countries are at a comparable level of development, which creates much greater competition. Taiwan, Hong Kong, Thailand, Singapore and the People's Republic of China are all formidable economic competitors in world markets. This greater competition is creating a wave of protectionist sentiment in the US and Europe which Japan did not face.

3. Korea shares a border with a hostile communist neighbor (north Korea) and must spend approximately 6% of GNP on defence. Military service in Korea is compulsory for all men. Japan does not require military service and is therefore able to direct more hours to productive activities. Japan spends slightly more than 1% of its GNP on defence. Japan's expenditures are tiny in comparison to the US's 6% of GNP and the 3.5% of GNP spent by NATO countries. Defence expenditure takes up nearly 28% of Korea's government budget versus only 6% in Japan.

4. Korea has no significant commercial experience upon which to draw. Major commercial activities began in Korea only in the 1960s, whereas Japan has been involved in international commercial trade for many centuries. Western influences came to Japan in the 1500s while Korea opened itself to the west only after the Second World War. Many of Korea's businessmen and technocrats are in their 30s and 40s and are just now gaining significant experience. Unlike their counterparts in Japan, these young managers do not have an older generation to learn from. They are exploring entirely new territory when they do business abroad and may make mistakes, although less constrained by established business practices.

5. Korea has not received the economic benefits of a nearby regional war. The Korean war was a huge boost to the Japanese economy as it recovered from the devastation of the Second World War. Japanese bases and industry played a key role in supplying the UN forces during the Korean war. The economic benefits of the Korean war allowed Japan's industries to grow to a level where their production exceeded pre-Second World War levels. While Korea benefited to some degree from the Vietnam war (Korea sent mercenary divisions to Vietnam), the economic benefit was not nearly as great. The economic contribution of the US military presence should not be underemphasized.

6. Korea does not yet have a stable political system. As is the case elsewhere in the world, political development often lags behind economic development. Japan's political maturity was accelerated when General MacArthur forced the country to adopt a stable civilian-dominated form of government. Korea has had to struggle to develop a democratic political system. This has been very difficult since Korea has a 5,000-year tradition of authoritarian political systems. Korea is only now beginning to develop a stable middle class capable of supporting a multi-party democratic system. For better or worse, General McArthur's imposition of a constitution and political system on Japan provided political stability at a critical time and a bridge from autocracy to democracy.

7. Korea has one of the largest foreign debt levels in the world. Korea's total foreign debt was more than US$35 billion at the end of 1987. In contrast, Japan was never a significant borrower from world capital markets even as a developing nation. In addition, Japan also had a positive current account balance much earlier than Korea.

8. A much larger portion of Korea's imports have been manufactured goods (including capital equipment) than was the case for Japan. At a comparable stage, Japan imported a much greater proportion of raw materials and commodities. This reflects the fact that Japan had a more significant machinery manufacturing industry at a similar stage of development.

9. Korea is also far behind Japan in its ability to produce and assimilate new technology. A study prepared by McKinsey & Co. found that in 1987 the number of Japanese engineers per 10,000 people engaged in work was 240. Japan had approximately the same number of engineers per 10,000 workers even when the per capita income of Japan was equal to present-day Korea (US$3,000). In 1983, Korea had only 32 engineers per 10,000 workers. The same ratio in the US is 160:10,000. Much of this lack of engineering talent in Korea comes from the brain drain. It has been reported that only approximately 40% of Korean students sent abroad for graduate study return to Korea after the completion of their studies. Most take jobs in the US which pay far more and which allow them greater responsibility at an early age. The Korean government has made an effort to repatriate this talent by offering incentives, but Korean engineers and scientists have not yet chosen to return in significant numbers.

10. In many respects Korea is ahead of Japan in terms of its integration into the world economic community. Korean businessmen are making a much greater effort to develop a group of individuals who are fully capable of dealing with the west. Many promising scholars and young businessmen are being sent abroad to learn from the world's finest universities and businesses. The English language skills of Koreans already far surpass those which exist in Japan. Since English is more and more the language of international business, Koreans are building a foundation to compete successfully with Japan. To their credit, Koreans have attempted to integrate new words (such as computer) into Korean in a manner which preserves the original English pronunciation while Japanese are compelled (like the French) to make it Japanese. The result is an increase in the ability of Koreans to converse with English speakers.

11. Koreans are also much more comfortable in dealing with foreigners. Both cultures consider themselves to be superior to the West, but Koreans are more willing to interact and come in contact with foreigners. On a Tokyo subway, a Japanese person will only sit next to a foreigner if it is the last seat available. In contrast, a foreigner on a subway in Seoul will have a difficult time preventing every Korean in the car from talking with him and asking him questions.

The preceding discussion has focused on the differences between Japan and Korea. Obviously, there are similarities as well. Both countries have similar economic resources, cultural traditions and are located in northeast Asia. Korea has in many respects patterned its industrial policy on the Japanese model. Certainly, Korea would like to catch-up and surpass Japan as an economic power. Korean political, legal and educational systems are also borrowed from Japan to a large extent (as much as Koreans hate to admit it). The influence from Japan's colonial domination of Korea before the Second World War is still significant. Most Koreans in their 50s and older speak Japanese since it was the language of instruction when they were in school. Ties between Korea and Japan are also strengthened by the large number of Koreans living in Japan. These Japanese-Koreans have contributed significant capital and expertise to Korea. For example, the Lotte chaebol is owned by a wealthy Korean who lives in Japan. Many commentators believe that the 1988 Olympic Games may provide the same economic springboard for Korea that the 1964 Tokyo Olympics did for Japan.

Deceptive outward appearances

The buildings which line Seoul's streets have the same polished granite faces and imposing dimensions which one can find in any large city in an industrialized country. The citizens of Seoul observe western standards of dress and it is rare to find a Korean man out of a western-style business suit. New cars clog the streets and a modern subway runs underneath the city. However, inside the buildings, clothing, automobiles and subways are men and women who are vastly different from their contemporaries in the west. Koreans not only differ in language and culture from foreigners: their values, attitudes, motivations and even logic are unique. In other words, the visible culture (which is readily grasped by foreigners) is radically different from the invisible culture. Understanding these differences is crucial to successfully doing business in Korea.

Foreigners are often deceived by the fact that the vast majority of their contacts are with relatively westernized Koreans. It should be emphasized that even very westernized Koreans have a totally Korean personality they will step back into whenever it is convenient. To understand the Korean people, a foreigner must disregard appearance and examine their values, culture, tradition and behaviour. Reading some of the many books which have been written on the Korean culture and behaviour will usefully increase understanding.

The average Korean businessman will spend his adult life working very long hours. All businesses, including banks, are open for most of Saturday, and Sunday is often spent

This new office development in Pusan illustrates the high density of land use in Korea's cities

doing business. Businessmen generally leave the office between 7.00 pm and 10.00 pm and usually proceed to a restaurant or bar with friends, rather than home to their wives and children. Three or four days of vacation are taken in the first two weeks of August. This is considered sufficient since recreational facilities are extremely limited. Trips abroad are for business rather than pleasure since government permission to vacation abroad is available only in narrowly defined circumstances.

Korea continues to be a male-dominated society. It is also a bi-polar society in which contact between men and women is formal and ritualized. In nearly all cases, the wife chosen by a senior Korean businessman has been selected from a small group suggested by his parents and married after a relatively short courtship. Marriage is an institution intended to promote economic well-being and the continuation of the family line rather than romantic interest. The joining of the two families is as important as the joining of the individuals. Children are produced almost immediately and it is rare for any but the extremely wealthy to have more than two children. The husband will typically turn his entire pay envelope over to his wife for management and will exist on an allowance given by his wife. Most of his allowance will be spent on eating and drinking with his friends during numerous late night outings. The wives of most businessmen find solace principally in the company of other wives and in devoting their lives to their children. Working after marriage is extremely rare for a middle or upper class Korean wife.

After discovering the existence of this unique social structure, the western businessman often assumes that it is gradually changing toward the way of life which exists in the west. This is true only to an extremely limited extent. As is the case elsewhere in the world, the advance of western technology in Korea has not been accompanied by a corresponding amount of social, cultural and behavioural change. The western businessman must understand and accept this difference and modify his business behaviour accordingly.

Immigration

To obtain a long-term entry visa, a company or individual in Korea will be required to sponsor the employment of the foreigner by providing a letter to Korean consular officials. Additional letters of recommendation can be helpful in obtaining an approval. Large Korean companies maintain good contacts with Korean government officials responsible for issuing visas and can be very helpful to foreigners going through this process.

Once an individual enters Korea on a long-term entry visa, he must apply for a residence certificate within 60 days of his arrival. The application for a residence certificate must be made at the relevant District Immigration Office. Holders of a residence certificate who intend to leave and return to Korea must also obtain a re-entry permit from the District Immigration Office.

Applications to extend an entry visa must be made before it expires. An extension application should be submitted to the District Immigration Office. Remaining in Korea after the expiration of a visa may result in the imposition of a substantial fine and difficulties in obtaining another entry visa in the future. If a person is deemed by the Korean government to have performed unlawful acts or engaged in improper behavior, he may also be prohibited from entering Korea. Immunization against communicable diseases is required only for individuals coming from infected areas.

Customs and excise

A verbal customs declaration is possible if no valuables are being brought into Korea. However, if valuables are being carried into Korea or if the declaration is for unaccompanied baggage, a written declaration form must be presented to Korean customs officials upon arrival at the port of entry.

Korean customs officials have been granted a great amount of discretion in enforcing customs laws. It is a good idea to be polite and give the official a business card to introduce yourself. In general, customs duty is more likely to be imposed if multiple pieces of the same item are carried into Korea. There is a separate customs line for visitors bringing in commercial samples, but they may need to be marked to avoid paying duty.

Instead of imposing duties, customs officials may be willing to enter the items in the visitor's passport. If the passport is marked in this manner, the person will be required to take the item with him or place it in a bonded area each time he leaves the country.

If imported goods are deemed by a Korean customs official to be for resale in Korea, the importation will not be permitted. Importation of certain items, prohibited under any special statute (such as firearms or other weapons), will not be permitted even if they are household goods.

The importation of commercial quantities of nearly all imported agricultural products (particularly packaged products) must obtain quarantine approval as part of the customs clearance process. Strict rules regarding plant importation forbid the importation of a very wide range of fresh fruits and vegetables. The importation of processed food products requires the approval of the Ministry of Health and Social Affairs.

The attitude of the Korean customs officials is best illustrated by posters seen in customs offices throughout the country urging officials to "conserve foreign exchange for national prosperity". The pervasive anti-import and anti-foreign bias means that delays and major stumbling blocks are possible in any given case. For example, the American Chamber of Commerce has reported that Korean customs officials claimed a foreign company must pay the following charges for three photographic slide programmes which had no commercial value:

Duty	0.50 x $354 x 3 =	$531.00
Warehouse charge	5 days x $2.50 =	12.50
		$543.50

The arbitrary implementation of Korean customs regulations and discretionary decision-making power afforded to customs officials has resulted in many complaints by foreign businessmen. These efforts have made imports difficult and expensive, but this must be understood as an expression of their national duty. Ultimately, change must come from the government.

Currency

The basic unit of currency in Korea is the won. Coin denominations are W1, W5, W10, W50, W100 and W500. Banknotes come in denominations of W500, W1,000, W5,000, and W10,000. Larger cheques issued by banks are also used for "cash" payments after a proper endorsement is made.

Foreign currency can be converted into won at banks and government-authorized money changing facilities. The exchange rate of the won into foreign currency is controlled by the Bank of Korea. The Bank of Korea has adopted a unitary floating rate system. The exchange rate is determined based upon a trade-weighted basket of currencies of Korea's major trading partners. The US is the dominant trading partner and therefore the dollar is the most significant component of the basket of currencies used in managing the exchange rate float. The exchange rate of the won is managed to achieve certain government policy objectives. The exchange rate has, until recently, reflected a gradual depreciation of the won against the US dollar to stimulate exports. One-time adjustments in the value of the won have been avoided whenever possible, to allow the economy to gradually adjust. The Korean currency had a value of slightly less than W800 for each US dollar at the end of 1987. The won is expected to significantly appreciate against foreign currencies in years to come before levelling off in the 1990s.

Major international credit cards including American Express, Visa, Master Card, JCB Card, Carte Blanche and Diners Club are accepted at large hotels, department stores and restaurants. Local credit cards include BC, Korea Express and KOCA. Some smaller shops may charge a premium (typically 3-4%) if a credit card is used in making a purchase.

Living conditions for foreigners

Housing suitable for foreigners is generally available in Korea, particularly in Seoul (where most foreigners live) and Pusan. However, housing costs are high by most standards. Rent for an unfurnished three bedroom apartment can range from US$2,500 to US$4,000 per month. Lessors may require that the rent for the entire lease term be payable in advance or that a large key money deposit (a type of refundable security deposit) be made. Rents for houses range from US$3,500 to US$7,500 per month (house rents are also typically payable in advance). Adequate collateral security should be obtained by a foreign lessee before entering a lease calling for advance rent or key money. Utilities are generally the responsibility of the lessee and because of the extremes in Korea's climate, will be approximately 30% of the rental amount. Except for rare cases, foreigners are not allowed to purchase residential property. In Seoul and Pusan there are excellent primary schools in which English, French and German are the languages of instruction. There are two major English language schools in Korea: the Seoul Foreign School and the Seoul International School.

Medical facilities and emergencies

A health emergency for a foreigner in Korea can be a traumatic experience. Many medical facilities exist in Korea, but Korean doctors and staff may not speak English. If an emergency occurs, a foreigner should call an organization in Seoul known as HART. HART operates a 24-hour hotline for emergency referral information. The telephone number is 707-8212, and the office is located in the International Clinic at 5-4 Hannam-dong, Yongsan-gu, Seoul (close to Itaewon). HART can provide an ambulance, a home or hotel visit by a doctor or nurse or an admission to a medical facility. Routine health related questions can also be answered. HART is a nonprofit organization whose funding comes from membership dues. Short-term visitors should not hesitate to call HART even if not a member.

The following hospitals are more suitable for foreigners (all are located in Seoul): Severance (Tel. 392-0161), Seoul National University Hospital (Tel. 762-5171), Korea National Medical Center (Tel. 265-9131) and Ewha Women's University Hospital (Tel. 762-5061).

Prescriptions are generally not required in Korea to obtain drugs from a pharmacy. Korean pharmacists may be helpful in giving advice on minor medical problems. Most major hotels have a pharmacy which employs English speaking personnel.

Chapter XI

Business structures for foreign companies

Properly structuring the Korean business presence of a foreign company is crucial from the operating, legal and tax perspective. Set out below is a discussion of the available strategies for entering the Korean market.

Importing products

Importing manufactured products into Korea is the least risky but most difficult alternative. Since most foreign companies operate through a branch office, the following discussion assumes that a branch will be the importer since different procedures apply to branches. The procedures for joint ventures and subsidiaries are similar but unique in certain respects.

Under Korean law only a licensed importer or exporter can apply for an import or export licence, and until 1981, branches were not permitted to obtain *trader's licences*. Therefore branches had to hire a licensed Korean importer or exporter and pay service charges ranging from 2-10% of the value of the imported goods. After 1981, under certain circumstances, foreign branches were allowed to obtain *special trader's licences*, allowing them to operate under their own names and accounts and save the service charges they would otherwise pay to licensed trading companies.

Branches of foreign companies are treated differently from Korean companies in the actual licence procedures. Korean companies, in order to be eligible for general trader's licences, must have paid-in capital which meets certain standards and must export a specified amount of goods from Korea. These standards change regularly so the amounts will not be specified here. However, the levels are high enough to disqualify all but very large companies. Companies which only import goods obviously cannot qualify. Foreign branches are exempted from these requirements, provided that they export goods from Korea designated by the Ministry of Trade and Industry (MTI), and import goods manufactured by the head office or *affiliated companies* for stock sales in Korea. Foreign companies which qualify under this category are granted a special trader's licence. The goods designated by the MTI as being *export permissible* are generally the items specified under its annual *Import-Export Periodic Notice*. The foreign goods which may be imported are generally limited to raw materials and equipment as well as parts and accessories for industrial use. Obtaining permission to act as a licensed trader for consumer and other finished goods is very difficult. In any event, substantial time and money are necessary to obtain a trader's licence.

An obstacle jeopardizing a foreign company's status as a licensed trader is that the MTI defines affiliated companies as those in which 30% or more of the shares are held by the corporation to which the branch belongs. This interpretation is severely limiting because it may exclude the parent company and even sister companies, and has been widely criticized by foreign businesses.

A licensed trader must obtain an import licence from the MTI for each shipment. The regulations under the Foreign Trade Law also require that when the trader applies for an import licence, a *sale of goods confirmation sheet* must be submitted. This document usually must be issued in Korea by a licensed issuer, but in some cases may be issued directly by the foreign seller. Both Koreans and foreigners commonly refer to this document as an *offer sheet* and to the issuer as an *offer agent*, although the document can be presented after the contract between the buyer and seller has been completed.

The actual authority of an offer agent appointed by the foreign seller may vary widely depending upon the authorization specifically granted. Some foreign companies give offer agents certain powers over the terms of the offer in addition to the authority to negotiate and solicit orders. The authority of an offer agent may be limited to simply conveying the offers made directly by the seller to the buyer under the agent's name in order to satisfy the Korean legal requirements.

There are several exceptions to the general requirement that an offer sheet be issued by a licensed offer agent in Korea. A foreign seller may issue the offer sheet if: (a) the goods imported are raw materials or equipment to be used for exports, (b) no locally licensed offer agent is handling the goods in question and this fact is certified by the Association of Foreign Trade Agents in Korea, or (c) if the MTI specifically consents. A foreign-issued offer sheet must also be confirmed by one of the following entities: (a) the competent government authority or Chamber of Commerce of the place of issue, (b) the Korean diplomatic or consular office of the place of issue, (c) a notary public in the state of issue, (d) the Korean telecommunications authorities if the offer sheet is sent by cable, or (e) a foreign exchange bank if the offer sheet was sent through the foreign exchange bank by telex.

An offer agent is normally paid a commission by the foreign seller each time the agent issues an offer sheet. The commission charged by offer agents may be substantial (generally 4-10% of the import price of the product). This cost can be avoided by using a Korean distributor who is a licensed offer agent.

The Korean branch of a foreign corporation or a foreign-owned company or wholly owned subsidiary may apply for an offer agent licence from the MTI. The applicant must submit evidence that it is authorized to present offer sheets on behalf of foreign suppliers (including its own corporate entity). An agreement with a foreign supplier is typically used to meet this requirement. The offer agent licence is granted at the discretion of the MTI which has recently become very restrictive in approving these applications.

Whether importation of the products into Korea will be possible will depend upon whether they are favourably classified under the Import-Export Periodic Notice. In general, if the product is not domestically available (which usually means high technology in the case of manufactured goods, or an unavailable natural resource or raw material), or is one of a few well-publicized exceptions allowed in response to pressure by foreign governments, importation will be allowed. The Korean government has a strong preference for import substitution (also called localization). That is, inefficient domestic producers will be protected even if Korean consumers must pay a higher price or accept reduced quality. For example, since farmers form a potent political force in Korea (as is the case in Japan) importers of agricultural goods will always face considerable opposition even if Korean consumers would benefit through lower food prices. This is true even if the product is not available in Korea, since the view is that a navel orange imported is a domestic apple or tangerine sale lost.

Both formal and informal pressure is placed on Korean citizens by the government to buy domestic products even if foreign products can be imported. This pressure takes many forms and is even incorporated into primary school textbooks. It is common for individuals who publicly consume foreign products or drive a foreign automobile to be visited by a tax collector. Duties on imported products are very high and in some cases may exceed 300%.

Localization and standards

Two of the major considerations in importing goods into Korea are localization and quality standards. Localization is a government effort aimed at building and utilizing existing domestic production capacity. Standards set the objective criteria for preventing the importation of foreign products. Every Korean government ministry can set standards for localization and performance requirements that the foreign company must meet in order to import its product. While top Korean government officials can speak in broad terms about liberalized importation and minimal import performance requirements, the relevant ministry (which often delegates authority to local trade groups) actually sets the performance requirements. Some Korean ministries (such as the Ministry of Health and Social Affairs) are notoriously protectionist in their application of laws and government policy. Never assume that your product can be imported into Korea until you have determined it falls within a well-known permissible category or you have obtained a written approval from the relevant ministry or a competent authority. Be prepared for a great deal of negotiation and delay if a remotely similar product to the one you want to import is currently available in Korea.

Most foreign companies which aim to import products must engage the services of a Korean distributor for their products. Under the Monopoly Regulation and Fair Trade Law, all import agency agreements and long-term import agreements with terms exceeding one year (with the exception of import agreements relating to raw materials, semi-finished materials and capital goods) must be reported to the Economic Planning Board (EPB) within 30 days of their execution. The EPB has issued guidelines (in a document document commonly known as *Notification 50*) which identify certain unjust collaborative acts and unfair business practices in international agreements and is specifically empowered to cancel an agreement or require amendment of its contents should the Fair Trade Office so decide.

The following practices have been identified as unfair business practices in distribution agreements subject to the Monopoly Regulation and Fair Trade Law:

– To oblige the distributor to purchase parts and related items from the foreign principal.

– To restrict resale outlets, resale price or the sales area of the imported products.

– To restrict the agent from handling products which compete with the imported products.

– To unreasonably restrict a business enterprise other than the distributor from importing products from a business enterprise other than the foreign principal.

– To impose on the distributor unreasonably restrictive conditions which are generally unacceptable in view of international contract practice.

Obviously, the last business practice is sufficiently broad to allow the Korean government to reject any proposed agreement, if it is deemed to be contrary to the national interest.

Another major problem faced by foreign companies is the Korean distribution system. As is the case in Japan, independent distributors play a major role in the economy. Several layers of distributors may exist between a manufacturer and the retail market. Complicating matters is the fact that some distributors may sell a portion of their inventory directly to the public. The distribution process is also complicated by the fact that most retail outlets are small family-run enterprises stocked with a wide range of goods. While the number of department stores is growing, many such stores are merely collections of individual merchants leasing space from the same landlord.

Most of Korea's import trade is purchased under letters of credit. Only certain raw materials can be financed in a way that allows a foreign supplier to extend credit to the purchaser from the time of shipment. The standard settlement methods permitted by the government do not allow *seller financing* and must be either through sight letters of credit or *documents against payment*. Thus, a Korean company must finance the full C&F (cost and freight) price of their purchase for a period of approximately two months (from shipment of order through customs clearance), and then for one to six months (depending on turnover) before receiving the purchase price from the customer. Since Korean commercial banks are extremely reluctant to finance this type of import trade because of the inherent risk, the importer either must finance the transaction himself, or rely on the illegal kerb money market and pay an exorbitant interest rate. Even in the unlikely event that a Korean bank will loan the necessary funds, the interest charge would be substantial.

High taxes and other charges on imports into Korea must be considered by a potential importer. For example, the following are the taxes and other charges imposed on wine and 32-bit microcomputer imports in 1987 as computed by the American Chamber of Commerce in Korea. It is important to note that the customs duty is imposed before other indirect taxes and imposts are assessed thus leveraging the amount of such taxes and imposts by the amount of the duty.

Wine

FOB price/bottle	$2.00
+ Transportation insurance	0.17
+ 100% duty	2.17
+ Liquor tax or 25%*	1.19
+ National Defence Tax	
– on tariff (2.5%) 0.05	
– on liquor tax (30%) 0.36	
+ Education tax (10%)**	0.12
+ Handling charge (10%)	0.22
+ Korean National Tourism Corp. Commission (5%)	0.31
+ Value Added Tax (10%)***	0.66
TOTAL	7.25

* Levied on the duty-paid price times 1.1 to reflect handling costs
** Levied on the liquor tax
*** Levied on the total of all the above.

Microcomputer

FOB price/unit	$11,814.00
+ Transportation insurance	200.00
	$12,014.00
+ 20% duty	2,402.80
	$14,416.80
+ 2.5% National Defence Tax	360.42
	$14,777.72
+ 10% VAT	1,477.22
	$16,254.94
+ Customs broker fee	35.00
+ Fee to import license holder	78.00
+ Insurance and handling 100.00	
+ Trade association charge 4.00	
	$16,471.94
+ 10% commission	1,647.19
Lowest possible wholesale cost	$18,119.13

Technology licensing

It is vital for a foreign licensor to chose a trustworthy Korean licensee and that an adequate return can be gained from the arrangement. It has been reported that 97% of technology licensing agreements approved in Korea provide for a royalty of less than 5% of net sales. A royalty rate in excess of 10% is almost never allowed by the Korean government. Thus, a careful analysis must be made in each case regarding profitability, given the fact that a Korean licensee may be a competitor after the licence expires. Unfortunately, the danger of creating a competitor cannot be avoided since other foreign companies are willing to licence the same or equivalent technology.

Since foreign exchange approval must always be granted before a Korean company can make a payment overseas, a technology licence requires prior government permission. The Korean government will examine each proposed technology licence to ensure that it is in the interest of both the Korean company and the nation. As a general rule, technology must be *sophisticated, suitable* for the nation's economy and *dissimilar* to domestically available technology. In addition, the technology license agreement must not contain *unfair* provisions or be a pretext for sales of raw materials or other products. Until recently, a trademark licence was not allowed unless it was accompanied by the transfer of appropriate technology. Technology licences with Korean companies may include running royalties and disclosure fees, and in addition, per diem payments for foreign technicians should they need to travel to Korea.

In the case of transfers of sophisticated technology on appropriate terms, certain tax benefits and exemptions may be available to a foreign licensor. These tax benefits are likely to be reduced in the years to come as Korea liberalizes its trade relationships in response to foreign pressure. As of the end of 1987, in certain cases, a five-year exemption of taxes on royalties could be obtained as well as an exemption from Korean income taxation on wages and benefits paid to foreign technicians.

Informal joint venture

This form of business presence is most effectively utilized by Japanese companies. In this relationship, the foreign company supplies a combination of equipment, technology, know-how, materials and capital to the Korean company. Mitsubishi has such a relationship with the Hyundai Motor Company. Mitsubishi holds only 15% of Hyundai Motor Company shares, but profits handsomely from the relationship. Many Japanese electronic companies have also entered into this type of arrangement to produce video cassette recorders. The key to success lies in transferring technology and designs which require that the Korean company buy high value-added parts only from the informal joint venture partner. For example, Mitsubishi's car designs require the purchase of transmission, motors and other key parts from Japan. Hyundai manufactures (or purchases from subcontractors) the simple parts, assembles the automobiles by combining the Korean and Japanese parts and both companies prosper. Korean video cassette recorder makers import over 90% of the value of the finished product from Japanese companies through these arrangements.

Joint venture or subsidiary

Whether a foreign company should choose a joint venture or a subsidiary is rarely an issue for a foreign company since the Korean government may require it to form a joint venture with a Korean company as a condition of approval. If a choice exists, the general feeling among foreign businessmen is that a Korean partner is helpful in the short-term (the first two to three years) but a hindrance in the long-term. In general, the author's experience has been that the subsidiary route is preferable. The wholly owned subsidiary of a foreign company not only has greater prestige and conveys a longer-term commitment to doing business in Korea, but also provides much greater operating flexibility.

Joining forces with a Korean company in a joint venture can provide the following advantages for a foreign company:

– Access to existing manufacturing facilities and distribution networks
– Access to domestic management personnel and production workers
– Access to domestic capital sources
– Increased access to government officials
– Increased consumer acceptance due to nationalist sentiment
– Access to permits which may no longer be granted
– Ease of withdrawal (selling to an existing partner is more expedient than liquidation or sale to a third party).

Among the disadvantages of a joint venture are the following:

– Sharing management control
– Disagreements over business strategy
– Loss of control over technology
– The creation of conflicts.

The primary problem with the joint venture approach is the fundamentally differing view of Koreans and foreigners concerning profits. Foreign companies are more interested in earnings which can increase the P/E ratio over the short-term in the home country. The Korean company is not interested in paying dividends and instead will want to reinvest any earnings in the business. In short, the Korean company is more interested in growth and market share than dividends. The Korean government encourages the reinvestment of any earnings since the overseas remittance of dividends reduces the available supply of foreign exchange. Another way of looking at this issue is to contrast the foreign view of the joint venture as a profit centre with the Korean view of the joint venture as an integral part of the company group.

The problem with deciding what to do with earnings is accentuated by the fact that the joint venture must rely on the Korean company for a number of different goods and services. Thus, it can be difficult (or easy depending upon your perspective) to adjust profits by adjusting intercompany charges.

Many foreign companies have been successful in avoiding (or at least reducing) the problems associated with a joint venture by utilizing a combination of royalty-bearing trademark and technology licensing agreements, management agreements and raw material supply agreements. These agreements allow earnings to be repatriated to the foreign company without declaring dividends. It is also useful for a portion of the production to be exported or raw materials to be supplied by the foreign company. This allows profits to be adjusted by making changes in transfer pricing.

At present, a foreign company is more likely to operate successfully through a subsidiary if it intends to sell a product which has clear technological advantages, is relatively non-labour intensive and which does not require a large sales force or distribution system. In contrast, if the product to be sold is a consumer good which has a marginal price advantage and which must be brought to market through the complex and multi-layered Korean distribution system, a joint venture with a Korean company may be a good idea. While not all products clearly fit within either category, each of the above factors must be considered. A wholly owned subsidiary may subcontract a Korean company to handle distribution services, but many foreign companies have been dissatisfied with their efforts. In most cases, the problem is that the Korean company has not been given sufficient profit incentive. The Korean distributor will encounter a certain amount of government harassment and public criticism if it works too hard to sell the products of a foreign company. Only a good financial return for the Korean distributor can compensate for this disincentive.

It is crucial that the objectives of the parties be consistent if a joint venture is to be success. Unfortunately, no matter how coincident the interests of two parties may be upon entering into a joint venture, the chance that this will continue decreases proportionately as time passes. Economic conditions will vary widely causing changes in resources, markets and objectives. As a result, creating a mechanism for winding up a joint venture is extremely important.

Nearly all investment by foreigners in Korea is governed by the Foreign Capital Inducement Law (FCIL). The FCIL requires that an approval be obtained from the Korean government before the investment is made. While it is theoretically permissible for a foreign investor to acquire shares previously issued to a Korean if prior approval for such purchase is secured pursuant to the Foreign Exchange Control Law, it is in practice almost impossible to secure such an approval. One exception to this rule is where a Korean company is involved in an existing joint venture pursuant to an FCIL approval and the Korean company is having financial problems. Foreign investors usually prefer to make their investment pursuant to an FCIL approval because of the tax and other benefits for which the joint venture (or subsidiary) and the foreign investor may qualify under certain circumstances.

If an industry is open to foreign investment, the foreign investor may apply for permission to acquire up to 100% of the equity of a Korean corporation. In practice, the Korean government will exercise considerable discretion with respect to the proportion of the equity a foreign investor will be permitted to acquire. Some high technology companies have recently been successful in obtaining approval for a wholly owned subsidiary. In addition, very large exporters have also obtained approval for a wholly owned subsidiary. Foreign companies such as Union Carbide, K-Mart, Litton, IBM and Du Pont have wholly owned Korean subsidiaries.

The minimum amount of foreign investment eligible for approval under the FCIL is US$100,000 in nearly all cases. There is no maximum amount that may be approved. However, very large investments will be scrutinized very carefully because of their relative impact on the economy. In order for a foreign company to invest in a Korean enterprise pursuant to the FCIL, an application for foreign investment must be submitted for review and approval. The Korean government will refer the application for review and comment to the relevant ministry. In the case of a joint venture, the Korean government will also refer the application to the Fair Trade Office of the Economic Planning Board to determine whether the agreement contains any provisions that contravene the Monopoly Regulation and Fair Trade Law. If any such unfair provisions are discovered, the Fair Trade Office will require the removal or modification of the relevant provisions. It is usually possible, through prior discussions with the appropriate government officials, to discover which provisions are likely to be objectionable. The foreign investor and the Korean investor may discuss the reasons for these provisions with the Fair Trade Office. In some cases the Fair Trade Office has been willing to change or revise its rulings after these informal discussions.

In 1984, the FCIL was amended to permit foreign investment in all industries except those identified in a *negative list* published by the Ministry of Finance in connection with its Guidelines for Foreign Investment. The negative list is based upon Korean Standard Industrial Classification (KSIC) categories established by the EPB. The negative list is reviewed periodically in order to gradually remove currently restricted industries. In addition to the negative list, the Korean government has issued guidelines establishing special criteria for investment applications in certain restricted industries.

Korean law has reserved certain industrial categories for *small- and medium-sized* enterprises on the grounds that they are not in a position to compete effectively with companies which are affiliated with large Korean industrial groups (chaebol). In addition, manufacturers of products in other KSIC categories are required by government regulations and policies to acquire certain parts and components by sub-contracting from small- and medium-sized enterprises.

Foreign investment in categories reserved for small- and medium-sized industries is subject to certain restrictions.

The foreign investment application must be accompanied by a copy of the project plan, the joint venture agreement or other documents, and documents establishing the identity of the individuals involved. These requirements change regularly so there is no point in being more specific here. Under the FCIL, any project which satisfies certain conditions will be eligible for an expedited approval procedure without referral to the relevant ministries.

The FCIL requires that the entire approved foreign investment amount should be received by the joint venture in Korea within 24 months from the approval date. Upon completion of the investment, and after importation of the capital goods purchased with foreign funds, the joint venture may be registered with the Ministry of Finance as a foreign invested enterprise (FIE). The FCIL guarantees the remittance of (i) dividends from a joint venture formed pursuant to an approved investment application; and (ii) proceeds received from the sale of shares of such a joint venture provided that a report of the sale has been confirmed by the Ministry of Finance. While the remittance of liquidation proceeds is not explicitly guaranteed by the FCIL, under current Ministry of Finance policy, as a practical matter, authorization for the remittance of such proceeds is routinely available. Remittance of other fees and expenses must be separately approved under the FCIL (for example, royalties under a technology inducement agreement) or under the Foreign Exchange Control Law.

In order to obtain confirmation from the Ministry of Finance of a sale of shares by the foreign shareholder to the Korean shareholder, the shares must be appraised by the Korea Appraisal Board or by one of several designated securities firms. The Ministry of Finance is unlikely to approve the remittance of an amount in excess of the value of the shares as appraised.

Choice of business entity

In addition to selecting the proper structure from an operational perspective, the proper legal structure should be chosen by the foreign businessman. The available legal forms of business presence in Korea are outlined below.

Liaison office

A liaison office is the lowest profile a foreign company can adopt when doing business in Korea. All the expenses of a liaison office must be reimbursed from abroad and the office may not generate any operating income. While the term liaison office is not a legal term, it is generally used to refer to an office which is engaged in nontaxable liaison activities of a preliminary and auxiliary nature to the principal business activities of the foreign company. The liaison office is not required to be registered with a Korean tax office as a tax presence of the foreign corporation or with the court as a branch. A nontaxable office may be utilized only for non-income-producing liaison activities which are conducted exclusively for the corporate legal entity of which it is a part. While the precise nature of the non-income-producing activities will vary depending upon whether there is a double-tax treaty between Korea and the relevant foreign country, in general they may include: (i) purchasing goods; (ii) storing or displaying goods which are not for sale; (iii) conducting advertising, public relations, collection and supply of information, market research, or other similar activities of a preliminary or auxiliary nature; (iv) having materials processed by others; and (v) other liaison activities for its home office and other branches of the same corporate entity. If non-income-producing activities are conducted for other entities, or if the office activities exceed the scope of the permitted liaison activities described above, the office must be registered as a branch or penalties may be imposed. It is common for tax officials to appear at company offices and seize communication records so as to determine whether tax free status is justified.

While a liaison office results in the lowest tax exposure, it can be difficult for a foreign company to utilize this type of presence since a liaison office has no official status. As a result, transactions must be conducted in the name of the liaison office manager. It may be difficult for him to purchase a car or engage in certain other activities which require an official registration in the company name. Some liaison offices are technically registered as a branch even though remaining nontaxable for that reason. A liaison office also has less prestige than a branch or Korean corporation.

Branch office

The term branch office appears in the Korean Commercial Code and refers to an office registered with district court. A branch office is usually taxable, but may be operated as nontaxable liaison office. While a branch may in theory undertake virtually any income-producing activity, certain activities are subject to specific restrictions under various Korean laws and regulations. The Foreign Exchange Control Regulations require that a foreign corporation wishing to establish a business presence (e.g., an office or a branch) in Korea must decide in advance what activities it will undertake. If the branch office intends to undertake income-producing activities and repatriate profits to its head office, it must obtain an approval from the Bank of Korea before its establishment. A branch is the most common type of entity used by foreign businesses in Korea. The Korean government considers branch applications on a case-by-case basis. The branch must maintain a legal representative who resides in Korea (either domestic or foreign) at all times. A branch will be taxed on Korean source income as if it were a domestic corporation.

Branches become eligible for remittance approval three years from the date of the Bank of Korea approval for branch establishment. However, the Foreign Exchange Control Regulations restrict the amount of net earnings repatriatable by the branch each year for the initial five years to 20% of the total operating funds induced into Korea. Any profits from business activities of the branch other than those specifically approved will not be repatriatable to the head office. Remittances above a certain amount to Korea must be approved in advance. The Korean authorities have informed foreign companies that the "business activities specifically approved" will be very narrowly construed.

The Korean government is extremely reluctant to allow branches to engage directly in manufacturing activities. Branches which will provide financial or other types of services require special approvals from the Ministry of Finance or other relevant ministry.

Corporation

A company formed under Korean law may be operated either as a joint venture company or as a wholly owned subsidiary. There are four types of company forms that may be chosen by the subsidiary or joint venture: the unlimited partnership (*hapmyung hoesa*), limited partnership (*hapja hoesa*), stock corporation (*chusik hoesa*), and limited liability company (*yuhan hoesa*). These four kinds of organizations are distinguished from one another by the degree to which the members are liable for the business activities and obligations of the company. The liability of a shareholder in a chusik hoesa is limited to the amount paid for the stock. The liability of a member of a yuhan hoesa is limited to the amount contributed to the company, but with several exceptions. The liability of a member under the partnership alternatives is unlimited.

While four kinds of companies exist, a foreign corporation cannot form a hapmyung hoesa or hapja hoesa as a joint venture or a subsidiary since Korean law prohibits a corporation from becoming a partner in a partnership under Korean law. Therefore, as a practical matter, the business forms available for foreign businesses are the chusik hoesa and yuhan hoesa. In practice, nearly all joint ventures in Korea have been structured as a chusik hosea. In fact, 90% of all Korean companies according to tax office records are chusik hoesa. Korean investors prefer the chusik hoesa form because it is considered more prestigious. The yuhan hoesa is typically used for small, family-run businesses. One major factor favouring the selection of the chusik hoesa for foreign investment is that a yuhan hoesa application raises unfamiliar aspects for Korean government officials to evaluate.

The procedure for forming a chusik hoesa is similar to that which must be followed in western countries when forming a company. However, numerous government approvals and registrations will be required. To establish a corporation, articles of incorporation must be drawn and notarized. Seven or more promoters are required for incorporation, none of whom need be Korean nationals. Under the Civil Code, their status as promoters lasts only until the corporation is registered when a board of directors assumes control of the company.

An application for incorporation and registration of the company must be submitted to a district court with the following documents:

– Articles of incorporation
– Application form for share subscription
– Certificate of payment for shares on deposit in bank
– Documents detailing issuance of shares
– Request of directors' and auditors' investigation
– Investment licence issued by the Ministry of Finance
– Minutes of the board of directors' or shareholders' meetings to elect a representative director
– Minutes of the inaugural general meeting with the promoters
– Report on election of directors and auditor
– Certificate of power of attorney, if registration is done by proxy
– Document certifying actual subscription of shares

After registration and payment of registration tax, equivalent to 2% of paid-in capital, the company is permitted to operate. It can enter into contracts, acquire rights and obligations, possess intangible property such as patents and copyrights, own real estate, establish commercial credit and undertake business transactions. Within 30 days after the start of business the company must apply to the district tax office for a business licence. This licence must be certified by government authorities twice a year.

The minimum capital necessary to form a Korean company is W50 million. Shelf or paper companies do not exist in Korea for this reason. It is generally not possible to restrict a holder's ability to transfer shares, but an action for damages can be maintained against a transferor who breaches a covenant not to transfer shares. While shareholders have control over the distribution of dividends, limits exist which restrict the ability of corporation to pay a dividend. This computation is based on a corporation's net assets, paid-in capital and capital surplus.

A chusik hoesa is managed by a board of directors and a representative director with the power to bind the company. Many foreign companies have therefore structured the joint venture so that there are joint representative directors whose signatures are both necessary to bind the company. If a company has more than one representative director and they do not have joint status, either can enter into a binding agreement. Each Korean company must elect an individual or group of individuals to serve as the *statutory auditor*. The statutory auditor must not be a director or officer or employee of the company. This position is separate and distinct from an independent auditor who may conduct an annual financial review.

Business case studies involving foreign companies

Foreign companies are not always successful in the Korean market because they have not properly investigated the unique business environment in Korea. What works in other markets may or may not work sufficiently in Korea due to the special characteristics and preferences of Korean consumers. At the very least, extensive market research should be conducted before significant decisions are made.

Set out below are examples of notable successes and failures in Korea. In general, the first company referred to in the discussion has made significant mistakes while the second company has been more successful.

Two food products companies

A foreign company entered into a joint venture with a major Korean food producer to manufacture and sell a processed food item. The foreign company held an 80% equity share and the Korean company a 20% equity share. A market study was not conducted in Korea since it was thought to be too expensive to hire a qualified marketing firm. Instead, consumption figures from other markets were *fudged* to predict consumer demand in Korea. Extensive manufacturing cost studies were performed since they could be done by the first company's employees at no cost (a false assumption). A very large factory was constructed and an expensive advertising campaign conducted by the joint venture. The Korean joint venture partner was assigned distribution responsibilities.

The response of Korean consumers to the product was poor, sales barely equalling advertising costs during the first year. Observers have postulated that the lack of success occurred because it was never clearly communicated to Korean consumers that the company was selling a food to be consumed as part of a meal rather than a snack. Whatever the reason for the lack of demand, distribution quickly became a nightmare as products began to go stale on the shelves, since merchants did not properly rotate the stock. Some observers believe that many of the problems arose because the Korean company was not properly motivated to sell the food product since it had only a 20% equity stake in the joint venture. Disputes over proper distribution methods soon resulted and the joint venturers fell out. An independent distributor was brought in to replace the Korean joint venture partner with limited success. The problem worsened when a second foreign company entered the already small Korean market for the food product with a lower priced product. The situation has been salvaged to a degree by the first company after it began using Korea as an Asian production centre.

The second company used the existing manufacturing facilities of its joint venture partner, held a lower equity stake to motivate the Korean side and *piggy-backed* on the first company's consumer education campaign. The second company has not been wildly profitable but has obtained a substantially higher return on equity. The second company is now well-positioned to take advantage of other opportunities.

Two athletic shoe companies

Korea is the largest manufacturer of athletic footwear in the world. Both of the American companies discussed here purchase huge amounts of athletic shoes from Korean companies for resale in the US and other major western markets. Both of these companies have been very profitable. The first athletic shoe company had a significant problem with counterfeit versions of their shoes which could be found in any small shoe store in Korea. Local residents, foreign tourists and US military personnel purchased large quantities of these shoes. To make matters worse, entire container loads of counterfeit shoes were shipped by pirates to foreign markets (particularly Japan and other Asian countries).

The second company had an insignificant problem with pirated versions of its athletic shoes. The key to the second athletic shoe company's success in preventing counterfeiting was the decision to market its shoes in Korea despite its small size and low return on capital investment, and to do so by granting an exclusive licence to a Korean shoe manufacturing company. While the royalties earned were small, the second company had successfully enlisted the aid of a local company to stop domestic piracy. The Korean licensee has done an excellent job of preventing pirating since it knows the market and has the muscle to prevent counterfeiting. The first American company mistakenly believed that it was not worth licensing their product in the relatively small Korean market. Realizing its mistake, the first company now has a Korean licensee.

Two computer companies

Independently, two American companies decided at approximately the same time to enter into manufacturing agreements with large Korean chaebol to produce IBM-compatible personal computers. Both American companies agreed to supply the necessary technology and sell the product in major foreign markets. The Korean companies manufactured the computers and were granted licensing rights to sell personal computers in Korea. The first American company chose to enter into an agreement with a Korean chaebol which while very profitable, has a corporate culture which emphasizes self-reliance. The subsidiary of the Korean chaebol was headed by the son of the founder. The son had no technical training and had never worked overseas. This chaebol had been unsuccessfully experimenting with manufacturing personal computers for some time. To make matters worse, the chaebol chosen by the first American company had little experience in working with foreign companies in developing a new product. The personal computers manufactured for the first company were produced months behind schedule. Soon after the products manufactured by the first company arrived in the US, they had to be returned to Korea for major alterations.

The second American company chose a Korean manufacturer with a good reputation and history of working with foreign companies. This Korean company had experience in making related electronic products but had never made personal computers. The second company was headed by a professional engineer with substantial overseas experience. He had to answer to the chairman of the second company only for results. The second company's products have been one of the major success stories on the personal computer market.

Dow Chemical Company

In 1969, Dow Chemical Company invested approximately US$7 million in a joint venture with a Korean company (Korea Pacific). The equity in the joint venture was divided equally between Dow and the Korean partner. The venture was formed to produce vinyl chloride monomer (a raw material used in producing PVC plastic) for sale in the domestic market. Dow also invested US$150 million in a chemical plant to be operated by a wholly owned subsidiary known as Dow Chemical Korea. Dow Chemical Korea produced chlorine which was then sold to the joint venture for use in making vinyl chloride monomer. Because of poor demand for chlorine, Dow Chemical Korea had a very substantial excess of manufacturing capacity and began to suffer large losses. Over US$40 million was lost by Dow Chemical Korea in the 1981-82 fiscal year alone. To make matters worse, the joint venture could purchase chlorine abroad at half the price which could be offered by Dow's wholly owned subsidiary. Dow proposed a merger of the subsidiary and the joint venture. The Koreans wanted to break the long-term supply contract between the joint venture and the Dow subsidiary.

The fate of the venture was spoiled in part by poor markets for the products of the joint venture. However, other significant problems were Dow's belief that it was not being properly compensated for its capital, technology and management expertise; and the Korean party's belief that it was being overcharged for raw materials. This pattern often occurs when one party is responsible for supplying raw materials for the venture under a long-term supply contract. Disputes between the directors resulted in an acrimonious parting of the joint venturers. In 1982, Dow sold all of its interest in the joint venture and the subsidiary for only US$60 million. Dow has recently returned to Korea on a much more limited scale and with considerable caution.

A Japanese automobile manufacturer

This Japanese automobile manufacturer entered into an informal manufacturing agreement with a large Korean chaebol. A minor equity stake was bought by the Japanese company in the chaebol and favourable terms were given on several significant loans. The Japanese company produced a design for a Korean automobile which found a niche in the market and which required that significant amounts of high value-added parts (i.e., engines, transmissions and electric motors) be purchased only from the Japanese company. Japanese technicians were dispatched to Korea and key employees of the Korean company were trained at company facilities in Japan. The key element lay in designing the automobiles so that only parts and components produced by the Japanese company could be incorporated in the final product. Some would say that the Japanese had established what is known as a *screwdriver* factory in Korea. In other words, the Japanese company supplied the parts and the Korean company supplied the screwdrivers to put them together into a finished product. In fact, some basic manufacturing was done by the Korean company but the highly profitable work was retained by the Japanese. The Japanese company also provided some sales assistance in foreign markets. The Korean product has been very successful and the Japanese company has profited handsomely from sales of parts and components as well as from an increase in the price of its shares in the Korean company.

Two US trading companies

The first company had experience in a number of other Asian countries and a small staff in Korea composed entirely of Korean nationals. Because of rising production costs in other countries, most notably Japan, the first company decided to substantially increase its activities in Korea. As a result, a decision was made to send an expatriate manager to Korea. The company decided to send a manager who had excellent academic and professional credentials but no substantial overseas experience. As soon as the foreign manager arrived in Korea his problems began. The staff was uncooperative and roadblocks appeared in each direction he turned. Complicating matters was the fact that the expatriate was a perfectionist and accustomed to dealing with a substantial support staff. Within a few months, the guerilla tactics of the staff and the problems associated with opening a new office had overwhelmed the young manager. When a regional manager arrived to check on the operation he soon discovered that the expatriate manager had become unable to conduct any business. The expatriate was sent home after a little more than four months in Korea. Over a year later, because of complaints from Korean manufacturers, it was learned that the Korean staff had been demanding kick-backs in return for placing orders. Their lack of cooperation with the manager had been intended to cause his removal from the start so that they could continue with this profitable but illegal activity. It was not until two years after the removal of the first expatriate manager that a competent replacement manager could be located. The opportunity for large profits during the period when the branch had no manager was lost forever.

The second trading company also decided it needed an expatriate manager in Korea and sent a branch manager with experience in another major Asian market. He was a generalist who was given full operating authority and told to terminate any employees which were impeding his progress in developing the branch's activities. He was self-sufficient and accustomed to doing things for himself rather than delegating. He was very patient but persistent in pursuing an objective and was given full assurance that he had a bright future with the second company upon his return home. The employee of the second company has been stationed in Korea for a number of years and the operation has been very profitable.

Two food manufacturing companies

The first company had excess manufacturing capacity in the US for its snack crackers so it decided that exporting some of its products to Korea would be a good idea. The first company paid huge amounts to a Washington DC lobbyist to influence members of Congress who in turn encouraged US trade negotiators to threaten their Korean counterparts with trade sanctions unless the snack cracker market was opened. After many years, as part of the ongoing trade liberalization, foreign-made snack crackers were finally allowed into Korea. The packaging of the first company's crackers was not changed for the Korean market. The product itself was unremarkable in terms of taste and was more expensive than domestic products. An advertising campaign was formulated to sell the product in large quantities to average consumers despite the fact that Koreans had long since learned to make a quality snack cracker.

The second company also decided to sell a food product to Korea. However, it chose to sell a premium quality chocolate-covered cracker in Korea. The packaging of the product was upgraded and the focus of the advertising campaign directed at high income consumers. The advertising campaign showed the second company's product being eaten after a glamorous dinner party by sophisticated consumers.

The first company's products have gone stale on the shelf while the second company has made an excellent profit. Other high quality products have since been introduced by the second company. The key to the second company's success was finding a profitable niche in the market rather than trying to move into an already crowded and competitive industry. European companies have been doing this successfully for many years, most notably with automobiles, machine tools and confectionery.

Conclusions

It is difficult to generalize about the most profitable approach to doing business in Korea. Direct foreign investment can be profitable, but can also be a disaster. Importing, technology licensing and technical cooperation can also result in profits or losses. However, it can be said that foreign companies which carefully investigate the Korean market and enter with well-defined objectives are nearly always the most successful. Companies which come to Korea looking to profit from both domestic markets (sales in Korea), and trading (sourcing products from Korea) will also have a considerably greater chance of being successful. By adopting this dual purpose strategy, risks are diversified and government support (because of the exports) is virtually assured.

Appendix

Information for visitors

Transportation

Many international airlines have direct and indirect flight services to Korea, landing at one of three international airports: Kimpo, Kimhae, and Cheju. Nearly 90% of all international flights to Korea arrive at Kimpo which serves Seoul. Foreigners may also arrive in Korea via a ferry from Japan or on a cruise ship (both of which arrive in the port of Pusan).

Most foreign businessmen travel in Korea by using taxis. For all but *hotel* taxis, the fare is calculated by a meter on the dashboard. The meter fare will vary depending upon whether the taxicab is a *regular* taxi, which is usually yellow, orange, blue or green or a *call taxi*, which is generally gold. Fares for a regular taxi are W600 for the first two kilometres and W50 for each additional 400 metres. A waiting charge will be added if the car moves at less than 15 km per hour. From 12.00 midnight to 4.00 am, a 20% surcharge will be added. Call taxis are larger and better heated (or cooled) and may be paged by telephone. Call taxis charge W1,000 for the first two kilometres and W100 for each additional 400 metres. Regular taxis are much more numerous than call taxis. Hotel taxis are essentially cars for hire and the rates are negotiated with the driver. Taxi drivers may or may not speak English. If the driver does not speak English, giving him the name of a major landmark (e.g., a hotel) and using hand signals will almost always get you to your destination. It is a good idea to obtain a card from the hotel which contains directions in Korean on how to reach the hotel. Tipping taxi drivers is not customary except when baggage is brought in the car.

One interesting Korean custom is called *hapsung* (taxi sharing). When taxis are scarce, taxi drivers will respond to calls from potential riders standing on the street, even if he already has a customer. The potential riders will shout out the district where they want to go. If it is near the destination of the existing customer, he may be picked up. A driver will usually ask a foreigner's permission before doing so. This custom is technically illegal, but continues as it supplements the driver's income. The additional passenger will pay a roughly calculated (and slightly reduced) fee.

If you are visiting a Korean company, transportation will often be provided by the company to and from their office or plant. If you are an especially valued customer or client, a car and driver may be provided to you for your entire visit. Since long taxi lines may be encountered, many foreign businessmen choose to hire a car and driver for the entire stay.

Foreign businessmen may find it convenient to use Seoul's impressive new subway system. Four subway lines crisscross Seoul and its suburbs from 5.00 am to midnight. Lines are identified by a colour system. Line 1 is red and runs through the centre of Seoul. Line 2 is green and runs in a circle connecting downtown with the suburbs. Line 3 is orange and runs from southeast to northwest. Line 4 is blue and runs from northeast to southwest. Tickets can be purchased from vendors and machines. The subway fare is W200 for one zone and W300 for two zones. A machine will punch the ticket as you enter and collect the ticket as you leave.

With some effort a foreigner can also learn to use Seoul's complex system of bus lines. There are two types of buses in Seoul. Violet-coloured buses are cheaper (W130), more crowded and stop more often. Green and beige buses are more expensive (W350), have more seats but fewer stops. The route number can be seen on the side of each bus. Since no English markings or signs exist, riding the bus is much more difficult than riding the subway.

One important fact that foreign visitors should realize about Korea is that few streets are named and there is no logical system for locating a street address. Houses and buildings are numbered in the order they are built and not in the order they are located on a street. 217-1 can appear next to 57-2. An address provides a person with a ward (ku or gu), and precinct (dong). If the road is named, the suffixes ro, no or lo will be used. Blocks on a street are broken down into plots (ka or ga). Because of this confusing address system, maps are nearly always included in invitations or promotional materials.

Korea has an extensive rail network. Information on rail travel can be obtained from the Korean National Railroad (tel. 392-0078). Many foreigners choose to ride the train from Seoul to Pusan. The best type of train to take is the Saemaul-ho Express which has air conditioning and a dining car. This express train also runs from Seoul to Mokpo, Seoul to Yosu and Seoul to Kyongju. Tickets may need to be purchased in advance from rail stations and travel agents. Long-distance buses are another method of travelling between cities. Express buses are air-conditioned and run regularly. Buses serve nearly every city and tickets may be purchased in advance.

Korean Air has until recently been the only domestic air carrier in Korea. Recently, the Kumho business group was given permission to form a second airline. Daily flights are available from Seoul to Cheju, Pusan, Taegu, Yosu,

Kwangju, Sokcho, Taejon, Ulsan, Masan and Chinju. Tickets may be purchased from Korean Air offices or travel agents. Information can be obtained in the Seoul area by calling (02) 751-7114. The Seoul-Cheju flight takes about one hour, the Seoul-Pusan flight 55 minutes and the Seoul-Taegu flight 40 minutes.

Restaurants and entertainment

There are an enormous number of restaurants to choose from in Korea, particularly if Korean food is desired, but foreigners must be as careful as anywhere else when choosing a restaurant. It is best to stick to established first-class Korean, Japanese, Chinese and western restaurants and avoid small establishments and street vendors. Although hygenic standards are improving, the hepatitis virus is prevalent, so it is wise to avoid certain foods such as shellfish.

Korean food

Korean food is an extremely varied fare characterized by the spices used and the extensive preparation required. The usual meal will consist of many dishes including soup, vegetables, meat, fish and fruit. Kimchi, Korea's most famous vegetable dish, is served at every meal. Kimchi is typically made of cabbage, turnips and cucumbers and can be tasted in thousands of varieties. Every Korean family has its own variation on the basic kimchi recipe. In the course of preparation, kimchi is fermented and seasoned heavily with red pepper and garlic. The dish was originally prepared in clay pots and buried in the yard. The kimchi preparation process, developed before the advent of refrigeration, preserved vegetables through the winter.

Plenty of garlic, scallions, and hot peppers typify Korean cooking. Popular dishes with foreigners include bulgogi (marinated beef strips cooked on a charcoal grill), kalbi (barbecued beef short ribs), shinsollo (a broth of meat, fish, bean curd, and vegetables served in a bowl containing a compartment for charcoal), bibimbap (rice with vegetables and red pepper topped with a fried egg) and naengmyon (a bowl of thin buckwheat noodles containing meat, chopped scallions, radish, cucumbers and pieces of egg).

Entertainment

Korea has a wide range of bars, nightclubs, discotheques, theatre restaurants and casinos. Major hotels provide the usual range of entertainment, including the most popular discotheques, but evenings spent in this manner can be very expensive. Less expensive and more typically Korean entertainment can be found in beer halls. A bar girl can be arranged who will pour drinks, make conversation, and dance with her customer. Those visitors who like to perform will enjoy visiting a *karoke* bar where they can sing at a microphone backed-up by recorded background music. A more refined (and more expensive) experience can be had at a *kisaeng* house. These luxurious establishments feature female companionship, high quality food and musical performers. The Korean government has also allowed a limited number of gambling casinos to operate.

Personal safety and crime

Foreigners (including women) can feel safe walking the streets of Seoul or any other Korean city or village. However, in the event of political unrest, street demonstrations must be avoided. They are considered internal matters at which a foreigner has no place (except the foreign press). There is very little violent crime in Korea and even less involving foreigners. Korea's low crime rate is in part due to the fact that policemen and military personnel are ever-present. The lack of violent crime in Korea is helped by laws and regulations which prohibit private citizens from owning firearms. The usual precautions against burglary and theft should be taken.

Other information useful to visitors

Drinking water. Many foreigners choose to drink bottled distilled or mineral water. Major hotels have their own purification systems and many guests consume the water from hotel taps without incident.

Electricity. Most electrical current in Korean cities is 100 volts. However, hotels will generally have both 100 and 220 volt outlets. Rural areas may use 220 volts.

Telephones. Two W10 coins are used in making a local call on orange public phones which disconnect after three minutes. Newer phones (which are usually black or gray), will accept higher denomination coins and do not automatically disconnect if a sufficient number of coins are inserted.

Newspapers and television. The Korea Herald and The Korea Times are published in English every day except Monday. AFKN television is an English language station operated by the US military.

Value-Added Tax and tipping. In Korea, a Value-Added Tax (VAT) is levied on most goods at a standard rate of 10%. VAT is also levied at a standard rate of 10% on services received in hotels, tourist facilities and major restaurants. Tips have not traditionally been a part of Korean customs, but 10% service charge is added to the bill at most hotels and other tourist facilities. Today even workers at traditional places frequented by foreigners have begun to expect tips.

Korean language. The hangul alphabet and phonetic writing system can be learned quickly by a foreigner with a little effort. Learning the nonphonetic writing system which uses

Chinese characters is a very difficult undertaking. Fortunately, most directional signs are phonetic and do not use Chinese characters. The hangul alphabet is phonetic and consists of 11 vowels (10 are now used) and 17 consonants (14 are now used). The Korean language itself is relatively difficult to learn for westerners, but has been mastered by many foreigners. The language is grammatically similar to Japanese but very different phonetically. The sentence order is subject, object, verb.

Reading romanized signs in Korea is complicated by the fact that two different systems are in use. The first is known as the Ministry of Education System, and the second the McCune-Reischauer system. The government has abandoned the Ministry of Education system in favor of a modified McCune-Reischauer system. Complicating matters further is the tendency of translators to adopt their own system. Thus, the same word can appear in several different romanized forms.

Also complicating the task of learning the language is the fact that word structure in the Korean language must change depending upon the relationship between the parties. A very honorific form must be used when speaking to a superior, a more informal form when speaking to a friend and a rough form when conversing with an inferior. Mastering the various levels (as well as determining who is superior), is a considerable task even if one can decipher the social ranking system.

Despite the difficulty which foreigners will have in learning Korean, mastering a few Korean survival phrases is easy and well worth the effort.

Business hours

Business and government offices are generally open from 9.00 am to 6.00 pm during the week and from 9.00 am to 1.00 pm on Saturdays. Government offices may close at 5.00 pm from November to February. Most professionals and businessmen can be reached for several hours after closing and it is very common for work to continue until late in the evening. Banks maintain the above hours except for a 4.00 pm closing time during the week, and a 1.30 pm closing time on Saturdays. Most retail stores are open from 10.00 am to 7.00 pm except for one day during the first part of the week. This day of closing varies depending upon the store. Many smaller establishments are open late into the evening. Foreign embassies and consulates are usually open from 9.00 am to 5.00 pm on weekdays and are closed on Saturday and Sunday.

Holidays

The following holidays are observed in Korea:

New Year's Holiday. The first three days of the Gregorian calendar year are celebrated. Many Koreans travel to their home towns during this period so little or no business can be accomplished. January 1-3.

Lunar New Year. Many Koreans, particularly in the countryside, observe the new year celebration based on the lunar calendar rather than the New Year's Holiday. This holiday is also known as Folklore Day. Variable (based upon a lunar calendar).

Independence Day. A celebration of the 1919 independence movement against colonial Japanese occupation. March 1.

Labor Day. Many banks and businesses offer this day off for employees. March 10.

Arbor Day. Trees are planted throughout Korea as part of the nation's reforestation drive. April 5.

Buddha's Birthday. A "feast of lanterns" highlights the birthday of Buddha. Colourful celebrations take place at Korea's temples. Variable (the eighth day of the fourth lunar month).

Children's Day. Programmes for children are featured on this holiday celebrated since 1975. May 5.

Memorial Day. This day is set aside to honour the nation's war dead. The largest ceremony takes place at the National Cemetery in southern Seoul. June 6.

Constitution Day. This holiday marks the signing of the Korean Constitution in 1948. July 17.

Liberation Day. This holiday celebrates the end of Japanese occupation in 1945. August 15.

Chusok (Thanksgiving). This day is a major holiday in Korea. It is traditional for people to visit their hometown to honour their parents and ancestors. The country literally shuts down during this period, so little business is done. Variable (the 15th day of the 8th lunar month and the day after).

Armed Forces Day. This holiday honours the nation's servicemen. It is noted for colourful military parades. Military units stage a major parade on Seoul's Yoido Island. October 1.

National Foundation Day. This celebration marks the day when Tangun, the legendary founder of Korea, is said to have established the nation in 2333 BC. October 3.

Hangul (Language Day). This day celebrates the anniversary of the promulgation of Hangul as the national language in 1443. October 9.

Christmas. This day is celebrated much like it is in the west by Korea's large Christian community. December 25.

Status of religions

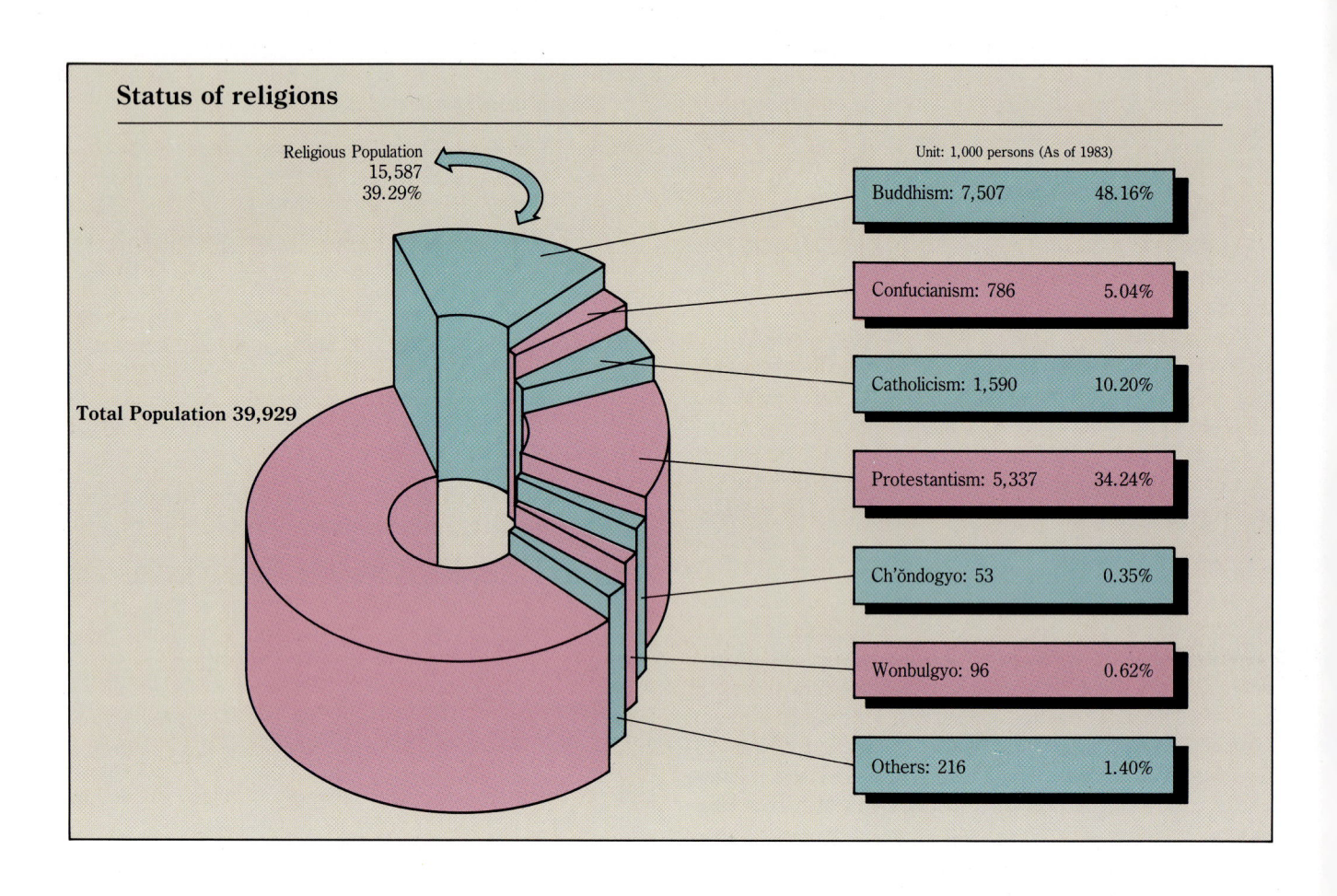

Religious Population
15,587
39.29%

Total Population 39,929

Unit: 1,000 persons (As of 1983)

Religion	Population	Percentage
Buddhism:	7,507	48.16%
Confucianism:	786	5.04%
Catholicism:	1,590	10.20%
Protestantism:	5,337	34.24%
Ch'ŏndogyo:	53	0.35%
Wonbulgyo:	96	0.62%
Others:	216	1.40%

Appendix

Business glossary and acronyms

Set out below is a list of some major government agencies, quasi-government organizations and business organizations (including their acronyms). The use of these acronyms is widespread in Korea.

AFDC:	Agriculture & Fisheries Development Corp.
AFKN:	Armed Forces Korea Network (US)
AKS:	Academy of Korean Studies
ALAL:	Alien Land Acquisition Law
AMCHAM:	American Chamber of Commerce in Korea
AMG:	American Military Government
BAI:	Board of Audit and Inspection
BOK:	Bank of Korea
CA:	Customs Administration
CBS:	Christian Broadcasting System
CFC:	Combined Forces Command
CSC:	Command Staff College
CSPI:	Composite Stock Price Index
CTL:	Corporate Tax Law
DBS:	Dong-A Broadcasting System
DJP:	Democratic Justice Party
DMZ:	Demilitarized Zone
EPB:	Economic Planning Board
E/L:	Export License
EUSA:	Eighth United States Army
FANSI:	Foreign Affairs and National Security Institute
FAS:	Foreign Agricultural Service (US)
FCIL:	Foreign Capital Inducement Law
FCL:	Foreign Exchange Control Law
ECL:	Foreign Exchange Control Law
FEEB:	Foreign Exchange Equalization Bonds
FIA:	Fisheries Administration
FKI:	Federation of Korean Industries
FKTU:	Federation of Korean Trade Unions
FTL:	Fair Trade Law
FTO:	Fair Trade Office
GSP:	Generalized System of Preferences
GTC:	General Trading Company
HS:	Harmonized System (Customs)
I/L:	Import Licence
JSC:	Joint Chiefs of Staff
JVA:	Joint Venture Ageement
KAA:	Korea America Association
KABI:	Korean American Business Institute
KAERI:	Korea Advanced Energy Research Institute
KAICA:	Korea Automobile Industries Cooperative Association
KAIST:	Korea Advanced Institute of Science & Technology
KAL:	Korean Air
KASA:	Korean Amateur Sports Association
KASD:	Korea Association of Securities Dealers
KBI:	Korea Bankers Institute
KBS:	Korea Broadcasting System
KCA:	Korea Coal Association
KCAB:	Korean Commercial Arbitration Board
KCC:	Korean Commercial Code
KCCI:	Korean Chamber of Commerce and Industry
KCGF:	Korea Credit Guarantee Fund
KDB:	Korea Development Bank
KDI:	Korea Development Institute
KEA:	Korea Electric Association
KEB:	Korea Exchange Bank
KECT:	Korea Emerging Companies Trust
KEDI:	Korean Education Development Institute
KEF:	Korea Employers Federation
KEIC:	Korea Export Industrial Corp.
KEMC:	Korea Energy Management Corp.
KEPCO:	Korea Electric Power Corp.
KESCO:	Korea Electric Safety Corp.
KERI:	Korea Economics Research Institute
KFB:	Korea Federation of Banks
KFTA:	Korea Foreign Trade Association
KGC:	Korea Gas Corporation
KGMP:	Korea Good Manufacturing Practice
KGT:	Korea Growth Trust
KHC:	Korea Highway Corp.
KHIC:	Korea Heavy Industry and Construction Corp.
KIET:	Korea Industrial Economy & Technology Institute
KIER:	Korea Institute of Energy and Resources
KIET:	Korea Institute of Economics and Technology
KIT:	Korea Institute of Technology
KIT:	Korea International Trust
KLDC:	Korea Land Development Corp.
KMPA:	Korea Maritime and Port Administration
KMPC:	Korea Mining Promotion Corp.

KNFC:	Korea Nuclear Fuel Corp.	MOT:	Ministry of Transportation
KNHC:	Korea National Housing Corp.	MSB:	Monetary Stabilization Bonds
KNR:	Korea National Railroad	MTI:	Ministry of Trade & Industry
KNTC:	Korea National Tourism Corp.	NA:	National Assembly
KOC:	Korean Olympic Committee	NACF:	National Agricultural Cooperatives Federation
KODCO:	Korea Overseas Development Corp.		
KOEX:	Korea Exhibition Center	NEPI:	National Environmental Protection Institute
KOPEC:	Korea Power Engineering Corp.	NIC:	Newly Industrializing Country
KOTRA:	Korea Trade Promotion Corp.	NIH:	National Institute of Health
KNTC:	Korea National Tourism Corp.	NHB:	National Housing Bond
KSA:	Korea Shipowners Association	NTA:	National Tax Administration
KSCT:	Korea Small Companies Trust	OBSE:	Office of Bank Supervision and Examination
KSDA:	Korea Securities Dealers Association	OCA:	Office of Customs Administration
KSFC:	Korea Securities Finance Corporation	OCAK:	Overseas Contractors Association of Korea
KSE:	Korea Stock Exchange	ONTA:	Office of National Tax Administration
KSI:	Korea Standards Institute	OPA:	Office of Patents Administration
KSRI:	Korea Standard Research Institute	OSROK:	Office of Supply Republic of Korea
KSSC:	Korea Securities Settlement Commission	PE:	Permanent Establishment
KT:	Korea Trust	POSCO:	Pohang Iron and Steel Company
KTA:	Korea Telecommunications Authority	PPD:	Party for the Promotion of Democracy
K-TAC:	Korea Technology Advancement Corp.	PPO:	Public Prosecutor's Office
KTB:	Korea Travel Bureau	RDP:	Reunification Democratic Party
KUSEC:	Korea-US Economic Council	ROK:	Republic of Korea
LMC:	Labour Management Council	ROKNRC:	Republic of Korea National Red Cross
LUA:	Labour Union Act	SEC:	Securities and Exchange Commission
MAC:	Military Armistice Commission	SIT:	Seoul International Trust
MB:	Monetary Board	SITRA:	Seoul International Trade Fair
MBC:	Munhwa Broadcasting Corp.	SLOOC:	Seoul Olympic Organizing Committee
MCI:	Ministry of Culture and Information	SNU:	Seoul National University
MFA:	Ministry of Foreign Affairs	SMIA:	Small and Medium Industries Association
MFNT:	Most Favored National Treatment	SMIPC:	Small and Medium Industry Promotion Corp.
MNC:	Multinational Corporation	ST:	Seoul Trust
MND:	Ministry of National Defence	TFCN:	Treaty of Friendship Commerce and Navigation
MOAF:	Ministry of Agriculture and Fisheries		
MOC:	Ministry of Construction	TIA:	Technology Inducement Agreement
MOC:	Ministry of Communications	TLA:	Technology Licence Ageement
MOE:	Ministry of Education	UNC:	United Nations Command
MOER:	Ministry of Energy and Resources	USIS:	United States Information Service
MOF:	Ministry of Finance	USFK:	United States Forces Korea
MOH:	Ministry of Health and Social Services	USTR:	United States Trade Representative
MOJ:	Ministry of Justice	VAA:	Veterans Administration Agency
MOL:	Ministry of Labour	VAT:	Value-Added Tax
MOS:	Ministry of Sports	WTCK:	World Trade Centre Korea
MOST:	Ministry of Science & Technology		

Appendix
Korean government offices

Economic Planning Board
1, Chungang-dong, Kwachon City,
Kyonggi Province
Tel: 503-7171

Ministry of Science and Technology
1, Chungang-dong, Kwachon City,
Kyonggi Province
Tel: 503-7171

Ministry of Foreign Affairs
77, Sejong-ro, Chongro-ku, Seoul
Tel: 738-9601

Ministry of Home Affairs
77, Sejong-ro, Chongro-ku, Seoul
Tel: 735-7411

Ministry of Finance
1, Chungang-dong, Kwachon City,
Kyonggi Province
Tel: 503-7171

Ministry of Justice
1, Chungang-dong, Kwachon City,
Kyonggi Province
Tel: 503-7171

Ministry of Agriculture and Fisheries
1, Chungang-dong, Kwachon City,
Kyonggi Province
Tel: 503-7171

Ministry of Trade and Industry
1, Chungang-dong, Kwachon City,
Kyonggi Province
Tel: 503-7171

Ministry of Culture and Information
82-1, Sejong-ro, Chongro-ku, Seoul
Tel: 730-5595

Ministry of Labour
1, Chungang-dong, Kwachon City,
Kyonggi Province
Tel: 503-7171

Ministry of Transportation
168 Bongnae-dong, Chung-ku,
Seoul
Tel: 392-9801

Ministry of Communications
100 Sejong-ro, Chongro-ku, Seoul
Tel: 750-2114

Ministry of Energy and Resources
1, Chungang-dong, Kwachon City,
Kyonggi Province
Tel: 503-7171

Appendix
Economic and trade associations

The American Chamber of Commerce in Korea
Chosun Hotel, 3rd Floor
87, Sogong-dong, Chung-ku
Seoul, Korea
James W. Booth – Executive Vice President
Tel: 752-3061
Telex: K23745 Chosun Attn: AMCHAM
A private non-profit association of business executives from the US, Korea and other countries concerned with trade and investment issues. AMCHAM is the principal voice of the foreign business community. The organization lobbies the Korean government, briefs visiting businessmen, organizes conferences and surveys wage and cost of living levels.

Korean Federation of Small Business
Yoido PO Box 1030
Seoul, Korea
Tel: 785-0010
Telex: K26500
This organization represents small- and medium-sized Korean businesses and lobbies the government on their behalf.

Korea Chamber of Commerce and Industry
CPO Box 25
45, 4-Ka Namdaemun-ro Chung-ku
Seoul, Korea
Tel: 757-0757
Telex: CHAMBER K25728
This is a group of Korean businessmen working to improve the business climate. This group is less dominated by the chaebol than the Federation of Korean Industries. The KCCI operates an Industrial Promotion Centre in Seoul.

The Federation of Korean Industries
CPO Box 6931
8-1 Yoido-dong, Youngdeungpo-ku,
Seoul, Korea
Tel: 783-0821
Telex: FEK015 25544
This is an organization representing Korea's very large businesses. A business should have annual sales exceeding W40 billion and spend more than W1 billion annually on research to qualify for FKI membership. The FKI currently has over 400 members.

Korean Foreign Traders Association
(World Trade Centre Building)
10-1 2-ka Hoehyung-dong,
Chung-ku, Seoul
Tel: 771-41
Telex: K24265
The KFTA is a non-profit organization composed of all licensed traders in Korea. The government has delegated a number of functions to the KFTA. This organization decides who may import and export goods.

The Korea Trade Centre
159, Samsong-dong
Kangnam-ku
Seoul, Korea
Tel: (02) 550-1114
This sprawling exhibition centre stages a large number of trade fairs and exhibitions. Also known as KOEX, this centre is located in southern Seoul. This recently completed facility is the largest of its kind in Korea. Over 550,000 square metres of floorspace house both special and permanent exhibitions. The complex also has a trade mart and several exhibition halls. A major deluxe hotel is part of the facility.

United States Department of State
Country Office for South Korea
Washington, DC 20520
Tel: (202) 632-7717
This office can provide information on political issues affecting business and trade.

International Trade Administration
Department of Commerce
Washington, DC 20230
Attn: Korea Desk Officer
Tel: (202) 377-2522
This office can provide information on doing business in Korea as well as information on trade and business issues.

US-Korea Economic Council
88 Morningside Drive
New York, New York 10027
Tel: (212) 749-4200
A non-profit organization which promotes US-Korea trade relations.

Korean-American Business Institute
Room 808, Paik Nam Building
188-3 Ulchiro, 10-ka
Seoul
A non-profit organization which fosters cross-cultural business understanding by providing consulting services and organizing an annual seminars on business topics.

Korea Design & Package Centre
CPO Box 732
138-8 Yeonkun-dong, Youngdeungpo-ku,
Seoul, Korea
Tel: 782-2205/9

Korea Trading Agent Association
45-14 Yoido-dong, Youngdeungpo-ku,
Seoul, Korea
Tel: 782-2205/9

Korean Commercial Arbitration Board
CPO Box 681 Seoul
(World Trade Centre Building)
10-1 2-ka, Hoehyung-dong, Chung-ku,
Seoul, Korea
Tel: 778-2631/5
A quasi-public organization which conducts arbitration proceedings under the Arbitration Law.

Korea Federation of Textile Industry
CPO Box 732, Seoul
(World Trade Centre Building)
10-1, 2-ka Hoehyun-dong, Chung-ku,
Seoul, Korea
Tel: 778-0821/3
Telex: K22677

Korea Export Association of Textiles
CPO Box 3303
(World Trade Centre Building)
10-1 2-ka, Hoehyun-dong, Chung-ku,
Seoul, Korea
Tel: 778-2411/5, 753-2072
Telex: K23697

Korea Garments and Knitwear Export Association
CPO Box 8790
(World Trade Centre Building)
10-1 2-ka, Hoehyun-dong, Chung-ku,
Seoul, Korea
Tel: 776-4121/3, 778-3591/2
Telex: K28435

Electronic Industries Association of Korea
648, Yeogsam-don, Gangnam-Ku
Seoul, Korea
Tel: 553-0941
Telex: K28999

Korea Auto Industries Cooperative Association
1638, Seocho-dong, Gangnam-Ku
Seoul, Korea
Tel: 587-0014
Telex: K22373

Korea Society for the Advancement of Machine Industry
Yoido PO Box 110
13-3 Yoido-dong, Yongdeungpro-ku,
Seoul, Korea
Tel: 782-5614
Telex: K25659

Korea Machine Tool Industry Assn.
Yoido PO Box 581
35-4 Yoido-dong, Yongdeungpo-ku,
Seoul, Korea
Tel: 782-5330
Telex: K25754

Korea Mould & Tool Industry Cooperative
13-31 Yoido-dong, Yongdeungpo-ku
Seoul, Korea
Tel: 783-1711/3
Telex: K25659

Korea Foundry and Forging Industrial Association
Yoido PO Box 709
43-3 Yoido-dong, Yongdeungpo-ku,
Seoul, Korea
Tel: 782-6877, 782-6994
Telex: K26500

Korean Association of Ships
Machinery and Equipment Manufacturers
12-5 Yoido-dong, Yongdeungpo-ku,
Seoul, Korea
Tel: 783-6952/4

Korea Footwear Exporters Association
256-13 Gongduk-dong, Mapo-ku
Seoul, Korea
Tel: 718-9531

Korea Tire Industrial Association
736-17 Yeoksam-dong, Kungnam-ku,
Seoul, Korea
Tel: 554-0174

Korean Plastic Goods Exporters Association
(World Trade Centre Building)
10-1 2-ka Hoehyun-dong, Chung-ku
Seoul, Korea
Tel: 755-2916/7

Korea Fishing Net Industrial Association
CPO Box 709
(World Trade Centre Building)
10-1 2-ka Hoehyun-dong, Chung-ku,
Seoul, Korea
Tel: 752-3837

Korea Petrochemical Industrial Association
68 Kuunji-dong, Chongro-ku,
Seoul, Korea
Tel: 313-3301/4

Korea Pharmaceutical Traders Association
CPO Box 7234
(World Trade Centre Building)
10-1 2-ka Hoehyun-dong, Chung-ku,
Seoul, Korea
Tel: 752-8481, 653-2434

Korea Foods Industry Association Inc.
36-3 Yoido-dong, Yongdeungpo-ku,
Seoul, Korea
Tel: 784-0088/9

Korea Canned Goods Export Association
(World Trade Centre Building)
10-1 2-ka Hoehyun-dong, Chung-ku,
Seoul, Korea
Tel: 752-0506, 755-6496

Korean Living & Fresh Fish Export Association
(World Trade Centre Building)
10-1 2-ka Hoehyun-dong, Chung-ku,
Seoul, Korea
Tel: 755-1351, 752-3408

Korea Frozen Seafoods Export Association
(World Trade Centre Building)
10-1 2-ka Hoehyun-dong, Chung-ku,
Seoul, Korea
Tel: 755-1744, 7427

Korean Wallpaper Exporters Association
(World Trade Centre Building)
10-1 2-ka Hoehyun-dong, Chung-ku,
Seoul, Korea
Tel: 753/5153

Korea Leather & Fur Export Association
(World Trade Centre Building)
10-1 2-ka Hoehyun-dong, Chung-ku,
Seoul, Korea
Tel: 778-3901/4

Korea Hair Goods Export Association
(World Trade Centre Building)
10-1 2-ka Hoehyun-dong, Chung-ku,
Seoul, Korea
Tel: 752-8487/9

Korea Metal Flatware Exporters' Association
CPO Box 62026-1 Sunhwa-dong, Chung-ku,
Seoul, Korea
Tel: 752-9577, 776-5876

Korea Toy Industry Cooperative
361-1, 2-Ka Hangango-ro, Yongsan-ku,
Seoul, Korea
Tel: 792-9505, 9818

Korea Optical Industry Cooperative
125-1, 4-ka Chungmu-ro, Chung-ku
Seoul, Korea
Tel: 266-4565, 261,1587

Korea Musical Instrument Industry Cooperative
50-6 2-ka Chungmu-ro, Chung-ku,
Seoul, Korea
Tel: 272-8978/9

Korea Sporting Goods Industry Cooperative
1 7-ka Ulchiro, Chung-ku,
Seoul, Korea
Tel: 252-7154/5

Korea Stationery Industry Cooperative
36-3 5-ka Chungmu-ru, Chung-ku
Seoul, Korea
Tel: 261-1207, 266-0417

Korea Ceramic Industry Cooperative
53--20 Daehyun-dong, Seodaemun-ku
Seoul, Korea
Tel: 363-0361/3

Appendix

KOTRA overseas offices

The Korea Trade Promotion Corporation (KOTRA) is a non-profit government- owned organization with responsibility for promoting the country's external trade. KOTRA manages over 70 Korea Trade Centres around the world in furtherance of its objective. This organization provides information on investment and trade opportunities. KOTRA can arrange meetings with Korean businessmen and can supply economic and business data on Korea.

North America

Chicago
Korea Trade Centre
111 East Wacker Drive, Suite 519
Chicago, Illinois 60601 USA
Tel: (312) 644-4323/4
Telex: 253005 KOTRA CGO

Dallas
Korea Trade Centre
PO Box 58023
World Trade Centre, Suite 155
2050 Stemmons Freeway
Dallas, Texas 65258-0023 USA
Tel: (214) 748-9341/2
Telex: 732343 KOTRA DAL

Los Angeles
Korea Trade Centre
700 South Flower Street
Suite 3220
Los Angeles CA 90017 USA
Tel: 627-9426/9
Telex: 674639 LSA

Miami
Korea Trade Centre
1 Biscayne Tower, Suite 1984
Miami, Florida 33131 USA
Tel: (305) 374-4648
Telex: 515186 KTC UR MIA

New York
Korea Trade Centre
460 Park Avenue, Suite 402
New York, New York 10022 USA
Tel: (212) 826-0900
Telex: 64904 KTC NYK

Washington (Liaison office)
Korea Trade Centre
1030 – 15th Street NW, Suite 752
Washington, DC 20005
Tel: (202) 333-2040
Telex: 289608 KTCW UR

Toronto
Korea Trade Centre
Suite 2802, Box 28
Cadillac Fairview Tower, 20 Queen St.
Toronto, Ontario, Canada
M5H 3R3
Tel: (416) 977-8784/6
Telex: 06-23426

Vancouver
Korea Trade Centre
Suite 1710, One Bentall Centre
505 Burrard Street
Vancouver, BC
Canada V7X 1M6
Tel: (604) 683-1820,(604) 687-7322
Telex: 04-54276 MOOGONG VCR

Latin America

Bogota
Korea Trade Centre
Carrera 7, No. 73-49
Seguros Colmena, Oficina 401
Bogota, DE Colombia
Tel.: 211-2615, 211-2910, 211-2648
Telex: 43189 KOTRA CO

Buenos Aires
Korea Trade Centre
Avenida Cordoba 462, Piso 19
Departmento C Capital Federal
Buenos Aires, Argentina
Tel: 312-2203, 2206, 5218
Telex: 17230 KOTRA AR

Caracas
Korea Trade Centre
PO Box 5368, Caracas 1010
Piso 4 Torre Banvenez, Av. Francisco
Solao Lopez Con Calle
Acueducto, Sabana Grande
Caracas, Venezuela
Tel: 729676, 729672, 729768
Telex: 21619 MUGNG VC

Guatemala
Korea Trade Office
Embassy of the Republic of Korea
16 Calle 3-38, Zone 10
Guatemala, Guatemala
Tel: 680302

Kingston
Korea Trade Office
5th Floor, Imperial Life Building
60 Knutsford Boulevard
New Kingston, Jamaica
PO Box 4r82, Kingston 6,
Jamaica
Tel: 92-92630/1
Telex: 2438 KOTRA JA

Lima
Korea Trade Centre
Av. Rivera Navarrete 451, 40
Piso-B San Isidro
Lima 27, Peru
PO Box L, 18-4971
Lima 18, Peru
Tel: 42-2834
Telex: 21182 PE KOTRA LM

Mexico
Korea Trade Centre
Paseo de la Reforma
No. 250-207 Col. Juarez
06600-Mexico, DF
Tel: 511-92-99, 514-54-57
Telex: 01774465 KTC ME

Montevideo
Korea Trade Office
Casilla de Correo 626
Montevideo, Uruguay
Tel: 90-8504, 90-0636
Telex: 6987 KTMVD UY

Panama
Korea Trade Centre
Edificio Banco Union 1 Piso
Local A, Ave. Samuel Lewis
Campo Legro Panama,
Republic of Panama

Quito
Korea Trade Centre
PO Box 4752-A, Quito
Av. Amazonas 477 Y Robles Edif.
Banco de los Andes 3,
Piso-Oficina 320, Quito, Ecuador
Tel: 230316, 525449
Telex: 2416 KOTRA ED

San Jose
Korea Trade Office
PO Box 151-500, San Jose
Edificio Centro Colon, Piso 4
Apartado 10151-1000, San Jose
Costa Rica
Tel: 334207, 331836
Telex: 3141 KOTRA CR

Santiago
Korea Trade Centre
PO Box 2136
Fidel Oteiza 1956-Piso 16-B
Providencia Santiago-Chile
Tel: 2515700, 2512521
Telex: 645207 MOOGONG CT

Santo Domingo
Korea Trade Office
Embassy of the Republic of Korea
Avenida Tiradentes 22
Santo Domingo
Dominican Republic
Tel: (809) 532-4314, 4315
Telex: (346) 4368 GKSD

Sao Paulo
Korea Trade Centre
Av. Paulista 1439, Jonj. 132
Bela Vista, Sao Paulo, SP, CEP
01311-Braxi
Tel: (011) 287-5726, 288-8521
Telex: 113418 ECOR BR

Europe

Amsterdam
Korea Trade Centre
Parnassustoren 13th Floor
Locatellikade 1, 1076 AZ
Amsterdam, The Netherlands
Tel: (020) 73055/6
Telex: 16368 KOTRA NL

Athens
Korea Trade Centre
Sina & Vissarionos 9, 7th Floor
Athens 135, Greece
Tel: 3626540, 3641567
Telex: 219596 KTCA GR

Brussels
Korea Trade Centre
World Trade Centre 1er, 2eme etage
Blvd. Emile Jacqmain 162
1210 Brussels, Belgium
Tel: (02) 218-5132, (02) 218-5499
Telex: 26256 KOTRA B

Copenhagen
Korea Trade Centre
Vester Voldgade 7-9 3 Sal 1552
Copenhagen V, Denmark
Tel: (01) 126658, 128039
Telex: 15291 KTC DK

Frankfurt
Korea Trade Centre
Mainzer Landstr. 27-31
6000 Frankfurt am Main-1
Federal Republic of Germany
Tel: (690) 236895/7
Telex: 416357 KOTRA D

Hamburg
Korea Trade Centre
Steindamm 71, 2000 Hamburg 1
Federal Republic of Germany
Tel: (040) 3803342/3
Telex: 2162400 KOTRA D

Helsinki
Korea Trade Centre
Kalevankatu 4, 00100
Helsinki 10, Finland
Tel: 641422
Telex: 122863 KOTRA SF

Istanbul
Korea Trade Centre
Mete Cad. No. 24/1
Kat-2, Taksim
Istanbul, Turkey
Tel: 1435075, 1498223
Telex: 24490 KOTC TR

Lisbon
Korea Trade Office
Avenida 5 de Outubro, 35, 3 DT
1000 Lisbon, Portugal
Tel: 573353
Telex: 14350 KOTRA P

London
Korea Trade Centre
4th Floor, 16/21 Sackville St.
London W1X 1DE,
United Kingdom
Tel: (01) 439-0501/3
Telex: 22375 KOTRA G

Madrid
Korea Trade Centre
Jose Lazaro Galdiano, 4-1
28036 Madrid, Spain
Tel: 457-5929, 457-5955
Telex: 44093 KTP CE

Milano
Korea Trade Centre
Via Large 31, 20122
Milano, Italia
Tel: (02) 876-806, (02) 874-422
Telex: 312522 KOTRA 1

Oslo
Korea Trade Centre
Daeleneggata 20, 0567
Oslo 5, Kingdom of Norway
Tel: (02) 382210, 382359
Telex: 77334 KOTRA N

Paris
Korea Trade Centre
16, Rue Hamelin 75116
Paris, France
Tel: 704-51-60
Telex: 610475 F

Stockholm
Korea Trade Centre
Dobelnsgatan 95
104 32 Stockholm, Sweden
PO Box 19052, Stockholm
Tel: (08) 347343, 347353
Telex: 12384 KOTRA

Vienna
Korea Trade Centre
Mariahilferstrasse 77-79, 1/5
(Generali Center, 5th floor)
A-1060, Wien, Austria
Tel: (0222) 963073-4
Telex: 134945 KOTRA A

Zurich
Korea Trade Centre
Leonhardshalde 21
8001 Zurich, Switzerland
Tel: (01) 252-3526, 3528, 3140
Telex: 816415 KTCZ

Asia and Oceania

Bangkok
Korea Trade Centre
GPO Box 1896 Bangkok
8th Floor, Kong Boonma Building
699 Silom Road (opposite Narai Hotel)
Bangkok, Thailand
Tel: 233-1322/3
Telex: 32335 MOOGONG TH

Colombo
Korea Trade Office
PO Box 965, Colombo
2nd Floor Unit No. 210
Liberty Plaza, 250 Duplication Road
Colombo 3, Sri Lanka
Tel: 574441, 574442
Telex: 21513 KOTRA CE

Dhaka
Korea Trade Centre
PO Box 2832, Dhaka
28, Gulshan Avenue (South)
Gulshan Model Town
Dhaka 12, Bangladesh
Tel: 603012, 604866
Telex: 642417 KTC BJ

Hong Kong
Korea Trade Centre
GPO Box 5573
Korea Centre Building, 2nd floor
119-121 Connaught Road, Central,
Hong Kong
Tel: 5-459500, 5-459786, 5-459509
Telex: 73497 KOTRA HX

Jakarta
Korea Trade Centre
PO Box 362/JKT
J1. Gatot Subroto No. 57-58
Jakarta, Indonesia
Tel: 511408, 514523, 515524, 510296 (ext. 52, 56)
Telex: 45188 MOOGONG IA

Karachi
Korea Trade Centre
Sasi Town, House No. 20
Civil Lines-10
Abdullah Haroon Rd. Karchi-4
Pakistan
Tel: 513491, 521863
Telex: 23687 KTC PK

Kuala Lumpur
Korea Trade Centre
Oriental Plaza, 10th floor
No. 3, Jalan Parry
Kuala Lumpur, Malaysia
Tel: 420756, 429939
Telex: 31191 KOTRA MA

New Delhi
Korea Trade Centre
c/o Embassy of the Republic of Korea
B-9/1-B, Vasant Vihar
New Delhi, 11057 India
Tel: 674782, 67344
Telex: 65162 KTRA IN

Osaka
Korea Trade Centre
5th floor Shin Shibagawa Building.
6-1, 4-Chome, Dosho-Machi
Higashi-ku, Osaka, Japan
Tel: (06) 231-1026/1027, 222-2296
Telex: KOTRA J64880

Tokyo
Korea Trade Centre
Yurakucho Building, No. 10-1
1-Chome Yurakucho, Chiyoda-ku
Tokyo, Japan
Tel: (03) 214-6951/2
Telex: J24393

Rangoon
Commercial Attache
Embassy of the Republic of Korea
PO Box 445 Rangoon
118, Boundary Road
Shwegonding PO
Rangoon, Burma
Tel: 32055
Telex: 21344 RGNKTC BM

Singapore
Korea Trade Centre
PO Box 421, Singapore 9008
No15-07, Hong Leong Bldg.
16 Raffles Quay, Singapore 0104
Tel: 2213055, 2213056
Telex: RS 23281 MOOGONG

Taipei
Korea Trade Centre
PO Box 1555 Taipei
Chien Hsin Building, 7th floor
72, Nanking E. Road Sec. 2
Taipei, Taiwan
Tel: (02) 5813030, 5813031
Telex: 21052 MOOGONG

Auckland
Korea Trade Centre
CPO Box 4007 Auckland
6th Floor, Sun Alliance House
40-44 Shortland Street
Auckland, New Zealand
Tel: 735-793, 735-792
Telex: 2818 MOOGONG NZ

Melbourne
Korea Trade Office
Suite No. EC 02C
World Trade Centre, Melbourne
Victoria 3005, Australia
Tel: (03) 6141733
Telex: AA 34085

Sydney
Korea Trade Centre
PO Box H69, Australia Square
Suite 3616, Australia Square
George Street, NSW
Sydney 2000, Australia
Tel: 27-3369, 27-2524, 27-5961
Telex: 25517 KOTRA AA

Middle East

Amman
Korea Trade Centre
The Embassy of Republic of Korea
PO Box 3471, Amman, Jordan
Tel: 42182
Telex: 642182 KOTRAMJO

Baghdad
Korea Trade Office
PO Box 6097
Al Mansur Baghdad
9d15/22/278 Hay Aljame'a
Jadriya Baghdad, Iraq.
Tel: 776-5496
Telex: 213271

Cairo
Korea Trade Centre
Consulate General of the Republic of Korea
PO Box 358 Dokki
Cairo, Egypt.
Tel: 716698, 715543
Telex: 92317 KOTRA UN

Casablanca
Korea Trade Centre
Tour Habous, Ave. des FAR
8eme Etage, Tour 14
Droite, Casablanca, Morocco
Tel: 314280, 314232
Telex: KOTRA 27636M

Dubai
Korea Trade Centre
PO Box 12859 Dubai, UAE
New Dnata Building (Airline Centre)
Third Floor, Office 9-A
Deira-Dubai, UAE
Tel: 220643, 223285
Telex: 46294 KOTRA EM

Jeddah
Korea Trade Centre
PO Box 4323 Jeddah, 21491
Saudi Arabia
Tel: (02) 6690031, 6690073
Telex: 400066 KOTRA SJ

Kuwait
Korea Trade Office
Embassy of the Republic of Korea
PO Box 20771
Safat, Kuwait
Tel: 814004, 849143
Telex: 22606 MOOGONG KT

Manama
Korea Trade Centre
PO Box 5337 Manama
Diplomat Tower Building, Room 403
Diplomatic Area, Manama
Bahrain
Tel: 259863, 234018
Telex: 8707 KOTRA BN

Muscat
Korea Trade Office
PO Box 4887
Ruwi-Muscat, Sultanate of Oman
Tel: 706459
Telex: 3483 KOTRA ON

Tehran
Korea Trade Centre
PO Box 11365-7877 Tehran
No. 190 Sepahbod Gharani Ave.
Assemi Building 4th floor, Tehran
The Islamic Republic of Iran
Tel: 89-6694, 89-9789
Telex: 212395 MGTN IR

Tripoli
Commercial Attache
Embassy of the Republic of Korea
PO Box 987 Tripoli, Libya
9-4, Jamal Abdul Einasser St.
Aburuquiba Building, 4th Floor
Tripoli, Libya
Tel: 34138
Telex: 20894 KOTRA LY, Attn KOTRA

Tunis
Korea Trade Office
7, Rue Tei-mour, El-menzh
Tunis, Tunisia
Tel: 234497, 231415
Telex: 12157 CORTV TN

Africa

Abidjan
Korea Trade Centre
BP 3429 Abidjan Avenue Jean Paul II
Abidjan-Plateau, Cote d'Ivoire Tel: 324581
Telex: 22105 KOTRA CI

Addis Abada
Korea Trade Centre
PO Box 5978
Addis Ababa, Ethiopia
Tel: 151853, 154306
Telex: 21240 KOTRA ADDIS

Dakar
Korea Trade Centre
BP 571, Dakar
Thiers Appartment C41
Place de l'independance X Rue
Dakar, Senegal
Tel: 21-00-74
Telex: 3163 KOTRABL SG

Douala
Korea Trade Office
BP 744 Doula
2eme etage, Immeuble
No. 11, Rue Prince de Galles
Akwa, Doula, Cameroun
Tel: 42 79 58
Telex: KOTRADLA 5284 KN

Kinshasa
Korea Trade Office
Juin BP 2975 KINSHASA 1
Immeuble Virunga-ler etage
Apt. No. 5 Boulevard du Trente
Republique de Zaire
Tel: 23687
Telex: 21407 KTC FIH ZR,
21412 GK KIN ZR. Attn KOTRA

Lagos
Office of the Commercial Attache,
Embassy of the Republic of Korea
GPO Box 1019 Lagos
Plot 1388A, Olosa St.
Victoria Island
Lagos, Nigeria
Tel: 611519
Telex: 22370 KECA NG

Nairobi
Korea Trade Centre
PO Box 40569 Nairobi
Finance House 7th Floor
Loita St., of Koinange St.
Nairobi, Kenya
Tel: 28928, 20458
Telex: 22360 MOOGONG

Appendix
Foreign embassies in Korea

Argentina
135-53, Itaewon-dong, Yongsan-ku
Seoul, Korea
Tel: 793-4062

Australia
5-7th floors, 58-1, 1-ka, Shinmun-ro
Chongro-ku
Seoul, Korea
Tel: 730-6491/2

Austria
Room 1913, 19th Floor, Kyobo Building
1, 1-ka, Chongro-ku
Seoul, Korea
Tel: 732-9071/2

Belgium
1-65, Dongbinggo-dong
Yongsan-ku
Seoul, Korea
Tel: 793-9611/2

Bolivia
1501 Garden Tower Building
98-78, Wooni-dong, Chongro-ku
Seoul, Korea
Tel: 742-7170

Brazil
New Korea Building
192-11, Ulchi-ro, Chung-ku
Seoul, Korea
Tel: 742-7170

Canada
Kolon Building.
45, Mookyo-dong, Chung-ku
Seoul, Korea
Tel: 776-4062

Chile
142-5, Itaewon-dong, Yongsan-ku
Seoul, Korea
Tel: 795-9519

Colombia
No. 125, Namsan Village
Itaewon-dong, Yongsan-ku
Seoul, Korea
Tel: 793-1301

Costa Rica
376, Itaewon 2-dong, Yongsan-ku
Seoul, Korea
Tel: 793-4280

Denmark
701, Namsong Building,
260-199, Itaewon-dong, Yongsan-ku
Seoul, Korea
Tel: 795-4187/9

Dominican Republic
133 Namsan Village
Itaewon-dong, Yongsan-ku
Seoul, Korea
Tel: 795-1850

Ecuador
133-20, Itaewon-dong, Yongsan-ku
Seoul, Korea
Tel: 795-1850

Finland
6th Floor, Kyobo Building
1-1, 1-ka, Chong-ro, Chongro-ku
Seoul, Korea
Tel: 732-6223

France
30, Hap-dong, Seodaemun-ku
Seoul, Korea
Tel: 362-5547/8

Gabon
Room 701 Kunchang Building
238-5, Nonhyon-dong, Kangnam-ku
Seoul, Korea
Tel: 548-9912

Germany
51-1, Namchang-dong, Chung-ku
Seoul, Korea
Tel: 779-3271

Guatemala
B-1116, Namsan Village
Itaewon-dong, Yongsan-ku
Seoul, Korea
Tel: 793/1319

Holy See
2, Koongjung-dong, Chongro-ku
Seoul, Korea
Tel: 736-5725

India
37-3, Hannam-dong, Yongsan-ku
Seoul, Korea
Tel: 798-4257

THE SPONSORS

Banque Indosuez

The roots of Banque Indosuez go back over a century with the founding of Banque de l'Indochine in 1875. The Bank then developed its activities throughout Asia by opening offices in Hong Kong (1894), Thailand (1897), Singapore (1905), Japan (1939) and Malaysia (1958). The Bank also worked to build up its presence in the Middle-East: with a Saudi Arabian branch set up in 1948, which became a Saudi institution in 1977 under the name of Al Bank Al Saudi Al Fransi.

In 1975, Banque de l'Indochine was merged with Banque de Suez et de l'Union des Mines, the influential banking subsidiary in France and Europe of the large Compagnie Financière dé Suez financial group.

In 1982, the Bank was nationalized along with its parent company, Compagnie Financière de Suez. At the end of October 1987, Banque Indosuez and its parent company were privatized.

Today, Banque Indosuez is one of France's foremost banking institutions. It has built a worldwide reputation as a business-oriented bank for French and foreign customers alike. Besides offering a complete range of specialized investment and commercial banking services, including real estate and leasing activities, Banque Indosuez's traditionally global outlook and widespread presence abroad have also combined to mark the institution as a major international bank.

An international network spanning 65 countries

Banque Indosuez has continually reinforced its longstanding and in-depth international network; it is active in 5 continents and 65 countries. In North America, Banque Indosuez has concentrated on sophisticated value added products such as futures and options in the Chicago markets and investment securities in New York and Montreal. Thanks to a conservative credit policy, the Bank has limited commitments in South America. In Europe, Banque Indosuez is preparing for a unified financial and commercial market in 1992. The Bank is reinforcing its capabilities in financial centres across the continent. It is the only French bank with offices in all five Nordic countries. In the Middle East, it is one of the most active foreign banks in Saudi Arabia, Yemen, Bahrain. Present in Asia for over a century, Banque Indosuez is the leading French bank in this region, where its activities range from traditional banking to new financial services.

Recent developments

Banque Indosuez has focussed its efforts on fully utilizing its human and technical resources in the development of sophisticated financial

The bank's trading room in Paris.

services and products geared to the needs of its worldwide clientele. The Bank's policy has particularly emphasized the accelerated development of its innovative financial-engineering expertise for corporate and institutional customers.

In 1987, Banque Indosuez expanded its brokerage activities through several strategic acquisitions. In March 1987, the Bank acquired W. I. Carr, a London based broker, particularly known for its Asian network and expertise. In November 1987, Banque Indosuez purchased a minority stake in the Paris stockbroker, Cheuvreux-de-Virieu, an initial move that will lead to a 92% majority holding by 1990. Cheuvreux de Virieu is the third largest broker in France and specialises in international brokerage activities. Banque Indosuez has also laid the groundwork for an international (24 hours) futures trading network with the acquisition of the Singapore based, HCB Futures Pte, and the creation of two autonomous subsidiaries specializing in futures and options on both the Paris and Chicago financial instrument markets.

Banque Indosuez has developed its know-how and placement capacity in international equities through the Suez, Compagnie Générale d'Electricité and Société Générale privatisations, in addition to offerings for private companies such as Sanofi and Eurotunnel. In both France and abroad, Banque Indosuez has reinforced its mergers and acquisitions activities which included for example, the purchase of a minority interest in Cap Gemini Sogeti, Europe's leading computer services group by Compagnie Financière de Suez or the merger of the household maintenance product divisions of Rhône Poulenc (France) and the Dutch company AKZO.

Banque Indosuez is playing a major role in large scale European projects such as Eurotunnel and Eurodisneyland and the Bank's expertise in project financing led to other achievements around the world: in Hong Kong, for example, the Bank is lead manager for the second motor and rail tunnel project.

1987 Consolidated highlights

Balance sheet total : Ffr 255 billion
Shareholders' equity : Ffr 8 billion
Long term resources : Ffr 27 billion
Customer deposits : Ffr 76 billion
Customer loans : Ffr 110 billion
Net income : Ffr 868 million

At the end of 1987, Banque Indosuez's total consolidated assets were Ffr 255 billion, a slightly lower figure than the previous year.

Consolidated net income rose from Ffr 828 million in 1986 to Ffr 868 million in 1987; after deducting the remuneration due to holders of perpetual non-voting securities, net income attributable to ordinary stockholders was Ffr 690 million against Ffr 793 million in 1986.

At December 31, 1987, consolidated stockholders equity amounted to Ffr 8.2 billion against Ffr 6.5 billion at the end of 1986. Nearly the whole increase (Ffr 1.5 billion out of Ffr 1.7 billion) was accounted for by the issuance of subordinated securities on the French market in January 1987.

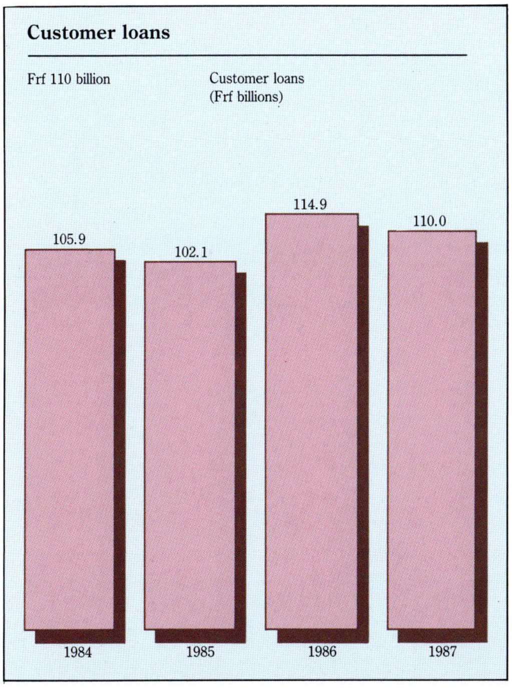

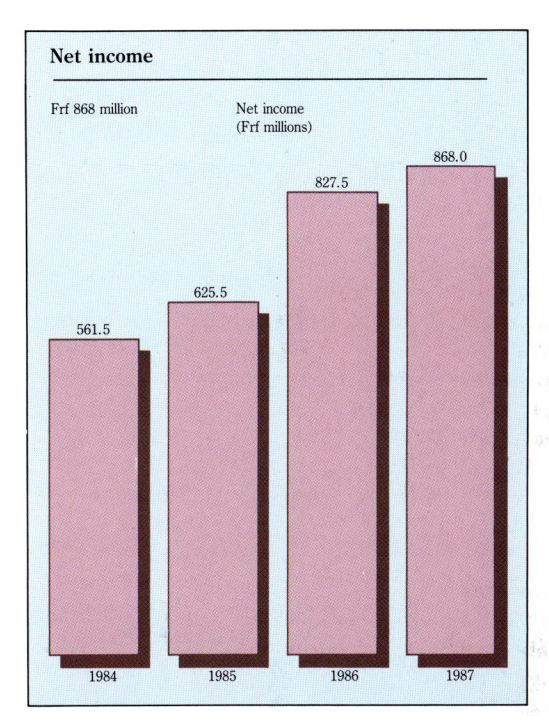

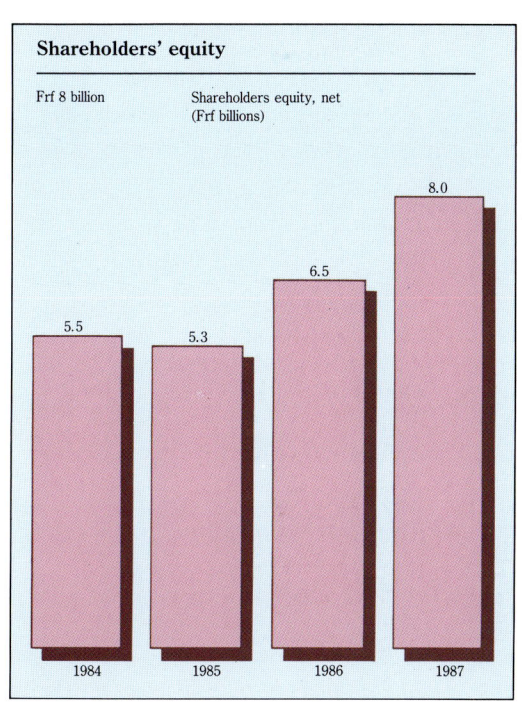

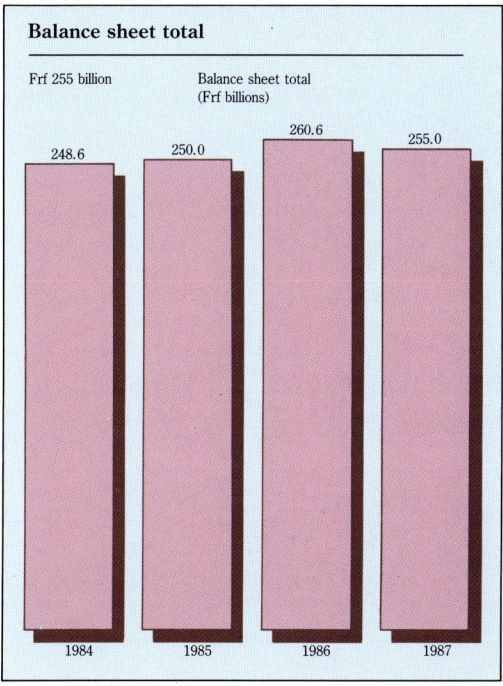

Customer loans
Frf 110 billion — Customer loans (Frf billions)

1984	1985	1986	1987
105.9	102.1	114.9	110.0

Net income
Frf 868 million — Net income (Frf millions)

1984	1985	1986	1987
561.5	625.5	827.5	868.0

Shareholders' equity
Frf 8 billion — Shareholders equity, net (Frf billions)

1984	1985	1986	1987
5.5	5.3	6.5	8.0

Balance sheet total
Frf 255 billion — Balance sheet total (Frf billions)

1984	1985	1986	1987
248.6	250.0	260.6	255.0

Banque Indosuez in Seoul

Banque Indosuez has a long tradition in Asia, dating back over a hundred years when it became the first French bank to be established on the continent. More recently, in 1974, it was the first French bank to open a branch in Korea.

Since then, the bank has grown from a team of 15, dealing with letters of credit and direct lending, to a full service branch of 45. The bank offers a wide range of sophisticated products to all of the major Korean industrial groups, placing the bank at the forefront of banking business.

Banque Indosuez, with its strong and diversified network in Asia and Europe, provides an excellent banking alternative to the American or Japanese partners. This reflects the more developed and internationally minded attitude of large Korean companies.

Mr JM Simart, General Manager

The general goals of Banque Indosuez in the world and Korea are:

1) To provide the most up-to-date products and services to our customers: Banque Indosuez maintains rigorously trained teams of highly qualified people to deal with traditional banking operations as well as innovative products. Regular contact with our valued customers is ensured by representatives from the specialised departments of the Bank as well as by its subsidiaries.

This is the case, for instance, of WICO, our representative office of W. I. Carr in Korea which in close cooperation with the Bank, proposes all our merchant banking services. On top of that, specific relationships with some carefully selected Korean partners, gives our customers access to domestic and international leasing and capital markets.

2) To provide worldwide support to large Korean groups which either want to tap the capital or financial market of a specific area or, to invest there, in line with their internationalization programme. Banque Indosuez's proven expertise and experience makes it a particularly reliable partner.

The best goals and the best policy are fruitless without the right people and the right team spirit. The recruitment, the training and the spirit of all the employees and officers have been, are, and will always be the first priority of the management team of the bank.

The management team, standing left to right: Mr CD Chung, Mr JS Park, Mrs JS Chae, Mr JM Simart, Mr CH Evain, Mr DW Lee; Seated: Mr JC Kim, Mr CS Kahng

Fully open to the future of Korea and its companies, the Seoul branch team brings to the latter more than 110 years of international expertise.

Mr CD Chung and his chief corporate dealer, Mr JC Kim, 'ready to face the next challenge'

Forex and Treasury

With the sizeable involvement of Korean groups in international trade, the dealing room is an essential part of our development.

Banque Indosuez has one of the largest dealing rooms of foreign banks in Seoul and covers all the products allowed by the regulators. The bank is already prepared to face the challenge of the market's liberalisation.

In close coordination with colleagues in Tokyo, Hong Kong and Singapore (and, Paris, New York, and London) our dealers quote all major currencies 15 hours a day and sometimes 24 hours a day.

Our experienced and highly trained people enable our clients to be certain that they are receiving the best advice possible.

Mr JS Park and his marketing team, 'an essential role'

Ranking 20th out of the 48 foreign banks in terms of total assets, but 16th in terms of profit, Banque Indosuez Seoul, is fully open to the future of Korea and its companies.

Marketing

Customer relationships remain the foundation of our development. Thanks to our 14 years in Korea and to the constant effort of our staff, we are able to offer an established and comprehensive service to our customers. They include the largest groups in Korea and several major European joint ventures.

The role of our marketing people is vital. They bring the Bank's services to the customers.

Mrs JS Chae and her team, 'dedication without limit'

Operations

In order to meet our customers' requirement for better services, our letters of credit team continually updates its knowledge of current regulations and, in full coordination with our worldwide network, matches the quality of our services to that of the best banks in Korea.

The dedication of the members of the team is without limit and is backed up by the most efficient electronic software and equipment available.

Mrs NL Sung and the electronic mail, 'the most efficient tools'

W. I. Carr

Local Expertise – Worldwide

W. I. Carr is a recognised expert in the stock markets of the Far East where it is a member of the Tokyo, Hong Kong and Kuala Lumpur exchanges. It is also active in a variety of other markets including UK equities and gilts, natural resources, Canadian and Continental equities. W. I. Carr traces its history back to 1825 when William Isaac Carr became a member of the London Stock Exchange. In March 1987 W. I. Carr was acquired by the Banque Indosuez Group, and through this affiliation is now part of a financial organisation that straddles East and West.

Hong Kong
W. I. Carr first came to Asia in the late sixties when it opened an office in Hong Kong in order to provide a facility for those clients who wished to deal in Far Eastern markets. In 1969 it became the first overseas associate member of the Hong Kong Stock Exchange.

Tokyo
Having established itself in Hong Kong, W. I. Carr then proceeded to form a steadily expanding network of offices in the region. The Tokyo representative office was opened in 1973, and in 1988 W. I. Carr was one of a select band of foreign broking houses to gain a seat on the Tokyo Stock Exchange.

Kuala Lumpur
In 1982 a representative office was established in Kuala Lumpur to provide coverage of the Malaysian and Singapore markets. In 1987, W. I. Carr was invited to become the first overseas corporate member of the Kuala Lumpur Stock Exchange. This enabled it to acquire 30% of the prominent Malaysian stockbroking house of Seagroatt and Campbell.

Seoul
Seoul was the next place in which W. I. Carr established an office, again at a time when very few foreign brokers were taking a serious interest in Korean financial markets. W. I. Carr's purchase in 1985 of a 5% stake in Daishin Securities Co Ltd sealed its commitment to Korea. The history and activities of the Seoul office are described in greater detail below.

Taipei
In 1987 a representative office was opened by W. I. Carr in Taipei, operating under a securities investment consulting licence. Its principal activity is the production of research by a small team of analysts, as well as other consulting functions.

Bangkok
The latest addition to the regional network has been the establishment in the summer of 1988 of an office in Bangkok to cover the Thai stockmarket, which unlike various other markets in the region, is open to direct foreign investment.

Research

W. I. Carr's Asian operations have always been research based. In each of the offices described above, including Korea, research teams have been built up to provide in-depth high quality research on the corporate sector, stockmarket, economy and politics. This expert local knowledge and analysis are the foundations upon which W. I. Carr's reputation has been built, and a wide variety of research products are distributed to clients internationally, on request.

Sales and dealing

Sales and dealing teams are situated in London, New York, Zurich, Hong Kong and Tokyo. Clients are serviced by experienced and well informed professionals who make extensive use of W. I. Carr's research and their own knowledge, and who maintain close links with overseas markets. The research, sales and marketing functions are supported by highly rated execution, settlement and administrative services, which utilise state of the art computer technology to ensure total efficiency and reliability.

Other products and sectors

Although this brief description of the W. I. Carr Group focuses mainly on its Asian expertise, and particularly on Korea, there are a number of other sectors and products in which the Group specialises.

- Natural Resources – This division of W. I. Carr specialises in providing research, sales and dealing coverage of oil and mining companies worldwide.
- North America – Capital Group Securities, with offices in Toronto, Montreal and Vancouver, specialises in Canadian equities, fixed income securities and corporate finance.
- Private Clients – W. I. Carr (Investments) Limited, based in England, incorporates the extensive private client business of Walter Walker (founded 1878) and Galloway and Pearson (founded 1867). A full range of stockbroking, nominee and valuation services are offered.
- UK Government Securities – W. I. Carr has long been recognised as an expert in this field, with a team of professional analysts and dealers concentrating on switching long-dated bonds using computer based analytical methods.
- Continental European Equities – The relationship with Banque Indosuez and French broking house Cheuvreux de Virieu place the Group in a unique position to provide coverage of the French and other European stockmarkets.
- UK Equities – W. I. Carr (UK) Limited provides a full research based service on the UK stockmarket to institutional clients.

W. I. Carr in Korea

W. I. Carr is a leading expert on the Korean securities market. Its representative office in Seoul provides research on the Korean securities market for foreign investors, complementing the efforts of sales teams in London, Hong Kong, Tokyo, and New York.

History
For many years, W. I. Carr followed the Korean market from its offices in Hong Kong and Tokyo. In 1985, W. I. Carr purchased 5% of the capital of Daishin Securities, a major Korean securities house, in order to increase its involvement in Korea. The following year, W. I. Carr opened a representative office in Seoul and quickly moved to establish its excellence in research. When the Korean securities market opens to more foreign participation, W. I. Carr will be ready to provide a comprehensive service to foreign investors.

Research products
The Seoul representative office now produces a daily market report, a monthly report, company reports, industry reports, and general information on the stock market and the Korean economy. The research focuses on the investment opportunities available to foreign investors through the existing trust funds and convertible bonds. The Seoul office has developed an extensive computer database of corporate financial information and stock market information. It also provides the current prices of the Korean trust funds and convertible bonds on Reuters (pages CARX and CARY).

Outward investment
W. I. Carr is eager to assist those Korean institutions that would like to invest abroad when allowed to do so. W. I. Carr distributes to Korean institutions research about overseas stock and bond markets, emphasizing its expertise in other Asian markets, European markets, natural resources, and fixed income instruments. The company has also been active in arranging overseas training in London, Hong Kong, and Toyko for a number of well-known Korean institutions.

Staff
The Seoul representative office is headed by Mr George W Long. Mr Peter Thorn is in charge of capital markets and outward investment activities. Mr Won-ki Lee and Mrs Jeong-ja Lee are investment analysts covering the Korean stock market.

Outlook for the Korean stock market
W. I. Carr is pleased to participate in one of the most dynamic economies and stock markets of the world. We are confident that the Korean economy will continue to grow in line with its historic average rate of real growth of 8% per year. That growth is now being reflected in the stock market, which has expanded dramatically over the past several years. We believe this expansion will continue and will ultimately mean that the Korean stock market, now capitalized at US$60 billion, will become the second largest stock market in the Asia/Pacific region and will be one of the ten largest in the world.

The staff of W.I. Carr's Seoul office.

W. I. Carr Offices

Seoul
W. I. Carr (Overseas) Ltd
Rm 2213 Kyobo Building
1–1 Chongro
Chongru-Ku
Seoul 110
Tel: (822) 7341250
Telex: K 34313
Fax: (822) 7320268

Tokyo
W. I. Carr (Overseas) Ltd
4th Floor
Yaesu Osaka Building
1–1 Kyobashi 1-Chome
Chuo-Ku
Tokyo 104
Tel: (81) 3 2784600
Telex: J222 6763
Fax: 03-278 9792

New York
W. I. Carr (America) Ltd
101 Rockefeller Plaza
New York
NY 10020-1903
Tel: (212) 3070511
Telex: 420402
Fax: 307 0648

Toronto
Capital Group Securities Ltd
141 Adelaide Street West
Suite 1810
Toronto
Ontario M5H 3L5
Tel: (416) 8607800
Telex: 6218236
Fax: 416-868 1524

Zurich
W. I. Carr
c/o Banque Indosuez
Bahnhofstrasse
18-Postfach 5127
Zurich 8022
Tel: (41) 1 2119464
Telex: 813705
Fax: 221 3457

Hong Kong
W. I. Carr (Far East) Ltd
21st Floor
St George's Building
2 Ice House Street
Hong Kong
Tel: (852) 5-255361
Telex: HK 73036
Fax: 5-8681524

Kuala Lumpur
W. I. Carr (Malaysia) Ltd
14th Floor, Menara Boustead
Jalan Raja Chulan
50200 Kuala Lumpur
Malaysia
Tel: (60) 3 2436411
Telex: MA 32424
Fax: 243 7513

Taipei
W. I. Carr (Taiwan) Ltd
Rm 1101 No. 270 Sec. 4
Chung Hsiao E. Road
Taipei
Taiwan
Tel: (886) 27716657
Telex: 16357
Fax: 781 9390

London
W. I. Carr (UK) Ltd
1 London Bridge
London SE1 9TJ
Tel: (441) 3787050
Telex: 8956121
Fax: 403 0755

Jersey
Le Masurier, James & Chinn
29 Broad Street
St Helier
Jersey
Channel Islands
Tel: (0534) 72825
Telex: 4192134
Fax: 345 9814

Daewoo Securities' Headquarters Building.

Daewoo Securities: the Korean capital market frontrunner

Extending its professional services and highly regarded investment expertise to clients at home and abroad, Daewoo Securities has always been at the forefront in developing Korea's capital market from its start-up operation in 1970. As the largest securities firm in Korea, Daewoo Securities provides a comprehensive range of financial services to financial institutions, corporations, and individuals through its diversified business components. Among the company's businesses are brokerage, underwriting, securities savings, corporate bond payment guaranteeing, and repurchase agreements.

Daewoo has developed a highly sophisticated and successful sales network in the domestic market. With an integrated and professionally trained salesforce of more than 2,000, Daewoo Securities' domestic sales network, which consists of 51 branches nationwide, is linked with the international market through its three representative offices in New York, London, and Tokyo. For years the company has consistently been a leader in the domestic arena. With paid-in capital of W114.1 billion (US$156 million) and market capitalization of W700 billion (US$914 million), Daewoo has enjoyed a dominant market share in each business field: 16% in equity trading, 15.9% in debt trading, and 10.5% in the underwriting of all types of Korean securities offered to foreign investors; straight bond offerings in the Japanese and European markets; international trust funds such as Korea Trust, Korea International Trust, Seoul Trust, Seoul International Trust, and Korean Growth Trust; the Korea fund and the Korea Europe Fund, close-ended funds listed in the New York and the London markets respectively; and euro-convertible bonds of domestic blue-chip companies.

Daewoo Securities is committed and well positioned to continue its expansion. Its future plans are based on the continued growth of the company's dominant domestic business and its successful entry into the international securities business. While Daewoo Securities has maintained its position as a market leader, it remains keen to cope with the challenges from other financial areas in international markets. In accordance with government guidelines, the internationalization of the Korean securities market has progressed smoothly, setting the final goal for full liberalization by the early 1990s. With this in mind, Daewoo Securities has expanded and strengthened its international operations to lead the internationalization effort.

Daewoo Securities' outstanding performance during the first stage which took place from 1981 to 1984 started out with the birth of the Korea Fund (US$60 million). Daewoo initiated the successful establishment of the fund with US prime firms like the First Boston Corporation in 1984. Additionally, in 1986, Daewoo Securities acted as co-lead manager in placing the second offering of the Korea Fund (US$40 million) which was also very successful.

In 1986, with government permission for Korean private companies to issue international securities such as convertible bonds or depositary receipts, Daewoo Securities co-lead managed along with Nomura and Goldman Sachs, the Euro-convertible bond issuance of Daewoo Heavy Industry Limited (US$40 million).

Chang Hee Kim – President of Daewoo Securities Co., Ltd.

Liberalization of Korean capital market

Since the inception of its international business operations in 1978, Daewoo Securities has placed itself on the forefront of Korean capital market liberalization. Korea has consistently pursued a completely open system for economic growth. Together with its partly liberalized trade and foreign exchange transactions, the government has regarded the globalization of the Korean capital market as one of the most effective means of escalating the Korean economy to match those of developed countries. In January 1981, the Ministry of Finance announced a four-step programme to liberalize the nation's capital market. the aim was to execute a gradual market opening specifically designed for Korea's economic conditions. The original plan was roughly as follows:

Stage I (1981-1984): Preparatory phase

- Establish international investment trusts and funds
- Grand opening of foreign securities firms' representative offices on the basis of reciprocity
- Set groundwork for internationalization by means of legal framework, computerized system, and personnel training

Stage II (around 1985): Developing but limited liberalization

- Allow limited portfolio investment by foreigners
- Permit Korean firms to issue and list their securities in global financial centres

Board of Directors: (left to right) Dong Yul Kwon, Executive Managing Director; Ki Sik Lee, Managing Director; Dong Il Cho, Executive Managing Director; Chang Hee Kim, President; Byoung Joo Ko, Standing Auditor; Ho Soo Lee, Director; Kun Whan Hahn, Executive Vice President.

Stage III (late 1980s): Further liberalization

- Allow direct foreign investment in domestic securities

Stage IV (early 1990s): Complete liberalization

- Fully liberalize the nation's capital market
- Permit domestic investors to participate in the international capital markets.

Taking its cues from government directives, Daewoo Securities has played a leading role by actively underwriting all kinds of Korean securities offered to foreign investors.

Korea Fund: a milestone for the liberalization of the Korean capital market

Outstanding among Daewoo Securities' active involvements was the co-sponsorship with First Boston to launch the Korea Fund (KF), a US$60 million closed-end fund listed on the New York Stock Exchange. The Fund was quite successful from its inception, with a good respose from foreign subscribers. The KF issue has since appreciated steeply from the issuing price of US$12 per share in 1984 to near US$80 at the end of March 1988. In May 1986, the Fund sold additional shares for the second tranche amounting to US$40 million to the investing public at the issuing price of US$32.5, in which Daewoo also participated as a sponsor of the underwriting group.

Among the six large domestic brokerage houses which currently receive transaction orders from the Korea Fund, Daewoo has handled the largest share of orders for the fourth consecutive year.

Since the initial public offering of the Korea Fund in 1984, Daewoo Securities has had its position extended in major markets. Backed up by the upward trend of the domestic stock market, the company co-lead managed the launching of convertible bonds for Daewoo Heavy Industries Limited in May 1986. The four equity-related securities which have been issued so far by Korean blue-chip firms were all launched successfully and have since been traded with high premiums. We expect that more Korean companies will tap into international markets with their equity-related securities starting in 1988. And Daewoo Securities will take the lead in introducing Korean blue-chip firms to the world financial markets.

Recently, in line with the current account surplus, Daewoo has extended itself to encompass underwriting close-ended country funds such as the Thai Fund, the Brazil fund, and the Thai Investment Fund.

Solid domestic operations

Brokerage/dealing

Over the past decade, Daewoo Securities has made a strenuous effort to maintain its leading position in equity and bond trading. During the fiscal year 1987, in the brokerage sector, the company handled W9.2 trillion, obtaining market share of 16.0% in equity trading and W2.7 trillion with 15.9% in bond trading respectively. These figures illustrate the excellent capability Daewoo has regarding brokerage functions and its current standing as the largest trading unit among 25 domestic securities firms. In the meantime, the market share by second securities firms for equity trading accounted for less than 9% during the same period. The accumulated knowledge and integrated service network using the most advanced computer systems enable Daewoo Securities to keep in close contact with its worldwide clients

Finally, in the dealing sector, the company has taken advantage of the highly favourable market climate. Accordingly, the company achieved huge gains amounting to W26.8 billion through dealing securities. Such a significant performance in the dealing business is largely attributed to Daewoo's arduous effort to attain the highest goal among competitors.

Underwriting

Daewoo has been a leading manager in the domestic underwriting business for the past decade and successfully maintained its position again in FY 1987 with equity issues of W55.9 billion, which accounted for a 22.9% market share and corporate bond issues of W362.8 billion with a 10.5% market share. This outstanding performance in terms of both total amount and market share, has reinforced Daewoo's reputation as Korea's most reliable financial intermediary for corporate finance.

Considering improved conditions and government encouragement for companies to go public, the company's underwriting business for equity and bonds will increase further in the years to come.

CP/corporate bond payment guaranteeing

Despite keen competition from other financial institutions with regard to the commercial paper underwriting and corporate bond payment guaranteeing business, Daewoo Securities' performance in these business areas is highly successful. The total volume of CPs underwritten and distributed by Daewoo for FY 1987 amounted to W29.9 billion and W20.2 billion respectively. During the same period, payment guaranteeing for corporate bonds amounted to W173 billion for 43 issues, amounting to a 20.8% market share. Such activities verify Daewoo's unwavering commitment to enhancing the quality and variety of its client services.

Securities savings

Daewoo Securities, for the first time, introduced securities savings in Korea to provide middle and low income people with a means of managing their assets. As a result, this move contributed to enlarging the stratum of investors in the Korean securities market. There are currently three different types of domestic securities savings: reserve, instalment, and employee-securities savings.

Other

Daewoo Securities has launched a new programme called the *bond management fund* (BMF). Since BMF carries many favourable features compared to financial vehicles provided by short-term finance companies and banks, it was relatively easy to attract huge amounts of capital. Among local competitors, Daewoo attained a 24.7% market share in this particular business by acccumulating around W300 billion as of March 31 1988.

Research and advisory services

For the Korea Fund's investments in Korean securities, Daewoo Research Institute, a Daewoo Securities wholly-owned subsidiary and the first investment advisory company of its kind in Korea, acts as the fund's only Korean adviser under the US Investment Adviser's Act of 1940. The Institute's fine research ability and investment advisory services have contributed greatly to the successful investment operations of the fund.

Staffed by 94 seasoned members in research and analysis, Daewoo Research Institute provides a full range of timely information about the Korean capital market and the Korean economy. For international investors, it has increased the number of periodic and non-periodic publications which cover a broad range of industries and individual companies. *Investment Survey of Listed Companies in the Korean Stock Market* and *Daewoo Securities Monthly* are among those publications widely recognized by investors worldwide. Meanwhile, for international investors, Daewoo Securities contributes daily information regarding Korean securities market movements to Reuter Monitor Service under the code name of DWSC through DWSM.

Another subsidiary firm wholly owned by Daewoo Securities is the Daewoo Capital Management Company (paid-in capital W2.2 billion, equivalent to US$3.0 million). Established in February 1988, the company is expected to provide valuable investment services for its clients both in and out of the country. In the midst of internationalization of the Korean securities market, Daewoo Securities will continue to expand and enhance its international operation in the years to come, and intends to remain the frontrunner in the Korean securities industry.

Looking to the future

Under a rapidly changing environment, Daewoo Securities stands firmly in the Korean capital market to realize continued growth. In 1987, Daewoo Securities experienced record profits due to prominent results from domestic business coupled with successful attempts to actively participate in international business. Although Daewoo Securities is proud of its past achievements, its principal focus continues to be on the future because Daewoo Securities' philosophy and responsibility are to lead the domestic securities market and to take a pioneering role in advancing into the international market. Having firmly positioned the firm in the competitive securities market, Daewoo Securities will continue to extend its business arena into the global market by focusing on future-oriented operations.

Looking to the future.

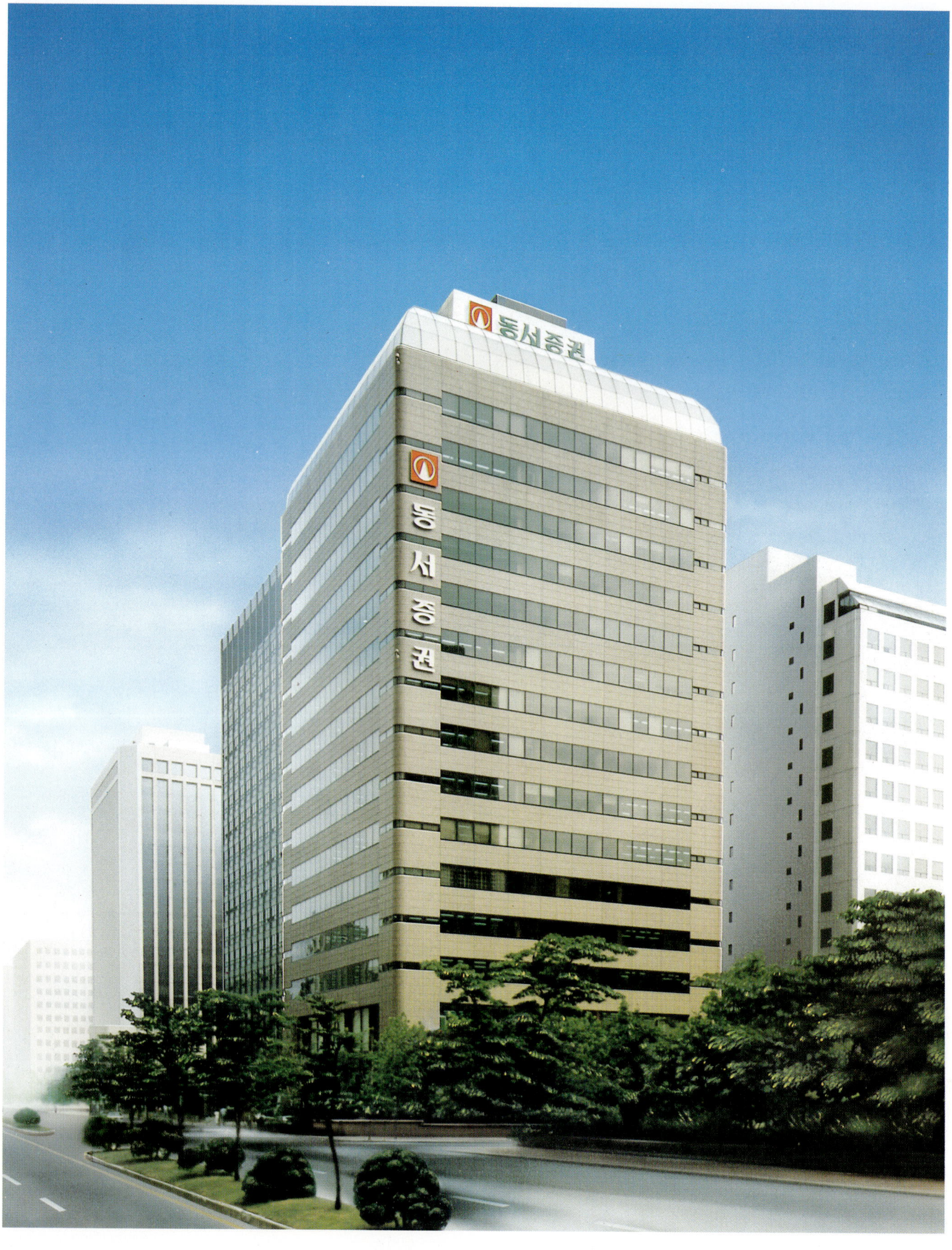
동서증권
동
서
증
권

Dongsuh Securities Co., Ltd.

Dongsuh Securities' Computer Information System, FOCUS, contains the most extensive investment information on the Korean market. Dongsuh is making every effort to furnish more valuable and useful information to its clients.

Head Office:
Dongsuh Securities Co., Ltd.
34–1, Yoido-dong, Yongdungpo-ku, Seoul, 150, Korea
Tel: 784-1211, 1411
Tlx: DONGSUH K24493
Fax: 784-2946, 5129

Overseas Representative Offices
Hong Kong Representative Office:
Suite 1503 Two Exchange Square 8 Connaught Place,
Central, Hong Kong
Tel; 5-8107686, 8107694
Tlx: 82720 DSHKC
Fax: 5-8450296

London Representative Office:
62 Mark Lane, London EC3R 7NE, United Kingdom
Tel; 481-1703
Tlx: 268949 DONGSU G
Fax: 481-1413

Tokyo Representative Office:
1–7, Kayaba-1 Chome, Nihonbashi, Chuo-ku,
Tokyo 103, Japan
Tel: 662-8924
Fax: 662-8925

Dongsuh Securities Co., Ltd. has gained an enviable position as one of Korea's foremost securities firms. Founded in 1953, Dongsuh has striven to achieve the highest level of professionalism by offering innovative financial packages to our clients.

In 1986, Dongsuh became a vital part of Kukdong Construction Co. Ltd, with a trained staff of over 1,300 represented through a network of 28 domestic branches, two subsidiaries, and three overseas representative offices located in London, Tokyo and Hong Kong. Dongsuh offers a full range of investment sevices including dealing, broking and underwriting as well as research and investment advisory services. The company also devises individually tailored financing for its corporate clients such as commercial paper brokerage, underwriting and sales, as well as bond guarantees and securities collateral loans. With the forthcoming capital market deregulation, Dongsuh stands prepared to take the next leap forward.

The advantages of investing in the Seoul market are becoming more obvious. Korea is now the seventeenth largest economy in the free world and the twelfth largest trading nation. Last year, exports rose 36%, real GNP growth was 12% and unemployment remained a low 2.2%. Equally impressive is Korea's securities market: the KCSPI rising to 68.9% in 1986 and 98.3% in 1987. With a market capitalization of W38 trillion (US$52 billion) and over 417 listed companies, the KSE is one of the most dynamic stock exchanges in the world today.

Dongsuh's position in the domestic market

With over 150,000 individual investor accounts, Dongsuh ranks second in the brokerage business and holds a market share of 9%. For the fiscal year 1987, brokerage commissions exceeded W34.1 billion (US$47 million), a sharp rise of 249%. Dongsuh is also an active dealer, playing an important role in stabilizing the Korean securities market. During FY1987, revenues from dealing totalled W37.86 billion, a rise of 276%. Among the five major securities firms permitted to guarantee corporate bonds, Dongsuh is the largest. With a 21% market share, the company is also the second biggest seller of commercial paper, which is permitted only to the six largest securities firms.

Staying ahead in today's competitive global market

Dongsuh has played a pivotal role in assisting corporate clients in obtaining financing from abroad. Dongsuh was the first Korean securities company to co-manage a convertible bond in the Euromarket when Samsung Electronics Co. issued a US$20 million

CB, the first of its kind from a Korean business group. In addition, Dongsuh played a co-manager role in the Yukong US$20 million CB, and an underwriting role in both the US$40 million Daewoo Heavy Industries and US$30 million Goldstar CB issues. Recently, Dongsuh co-managed the Korea Development Bank's US$20 million FRCD in the Euromarket. Very soon, Dongsuh will enter the US market when it opens a New York representative office. Dongsuh also plans to expand its Hong Kong representative office to a full branch or subsidiary.

Dongsuh has been vigorously encouraging foreign investment into the Seoul market. Dongsuh served as underwriter and broker for the Korea Fund Inc., whose shares are listed on the New York Stock Exchange and as Fund Manager and broker for the Korea Europe Fund Ltd, whose shares are listed on the London Stock Exchange. Dongsuh has an 18% equity share of the Korea Schroder Fund Management Co. Ltd, which manages the Korea Europe Fund.

Dongsuh has recently witnessed the growth of other newly industrialized countries, especially Taiwan and Thailand, and has actively been involved in the financing of investment fund companies in those countries. Dongsuh has served as either underwriter or part of the selling group of the Thai Fund Inc., Thai Investment Fund Inc., Thailand Growth Fund Ltd, and the Taiwan Fund.

Dongsuh's focus on information technology

Dongsuh has invested heavily in the improvement of its electronic data processing system. The system known as DONGSUH FOCUS is the largest EDPS of its kind in Korea and covers 1,400 different categories of information.

The services provided through DONGSUH FOCUS have won favourable recognition from clients, academic institutions, and newspapers. Munwha Broadcasting Corporation (MBC), Dong-Ah Ilbo and the Korea Economic Daily are constantly depending on DONGSUH FOCUS for up-to-the-minute financial information.

Dongsuh Securities Co is planning to network DONGSUH FOCUS into a Home Trading System and an Audio Response System to make it even more convenient for their busy clients.

Dongsuh's research into the future

The Dongsuh Investment Research Institute, established in 1986, is a wholly owned subsidiary with a staff of 50 economists and analysts. The Institute is engaged in the ongoing research and investment analysis of leading economic indicators, listed companies, industrial sectors, and stock market movements. The Institute is also engaged in software development to be used in conjunction with DONGSUH FOCUS and serves as a consultant to various regulatory agencies.

Dongsuh Investment Research Institute publications include the weekly Dongsuh Investment Guide, Securities Analysis Monthly, the quarterly Dongsuh Chartroom, the Investment Analysis of Listed Companies, Theory and Practice of Securities Investment, and the Dongsuh Monthly. These publications are available to clients and are also sent to over 500 institutional investors worldwide on a subscription basis.

More than 500 international delegates attended the International Symposium sponsored by Dongsuh Securities on May 25 1988. The subjects covered included the current situation of the Korean economy and the investment management company's role.

Investment management for Korea

Dongsuh Investment Management Co. Ltd, founded in the first quarter of 1988, is one of the first companies to provide investment advisory and management services to individual Korean and institutional clients. Previously, the company was part of the Dongsuh Investment Research Institute, but as demand for such services grew, a separate subsidiary had to be formed. With the average household income approaching that of advanced countries, a high savings rate, and government policies restricting real estate investment, Dongsuh Investment Management Co. has implemented a number of strategies designed to meet the growing needs of individual clients. The minimum investment amount is W100 million (US$140,000) and the minimum contract term is one year.

Behind the recent success of Dongsuh Investment Management Co. is a highly trained and experienced group of professionals. The researchers and fund managers that make up the compny have advanced degrees, including MBAs and PhDs in accounting, finance and economics, and a minimum of five years' securities experience. Working as Fund Manager for the Korea Europe Fund Ltd through the Korea Schroder Fund Management Co. Ltd has given Dongsuh experience in managing large-scale institutional portfolios.

Because this is a relatively new field in Korea, the growth potential is almost limitless. With the recent foreign exchange liberalization, Dongsuh Investment Management Co. will soon include overseas securities as part of its portfolio recommendations.

The reception after the international symposium on the Korean economy and securities market, sponsored by Dongsuh Securities. First from the left is Dai-Chung Kim, Executive Vice President, third from left is In-Kie Hong, President of Dongsuh Securities Co., and at far right is Kwan-Chong Kim, Vice President of Dongsuh Investment Research Institute.

The future role of Dongsuh Securities Co., Ltd.

Dongsuh hopes to expand its position in Korea to become one of the leading global financial institutions, in line with the growth of Korea's economy and domestic securities deregulation. Dongsuh will continue to strive for excellence in client satisfaction through the highest priority placed on human resources, employee recruitment and training. Dongsuh welcomes the participation of overseas investors in the Korean capital market either indirectly through the various fund companies, or directly when deregulation is completed.

Securities dealing operations through our excellent computer system.

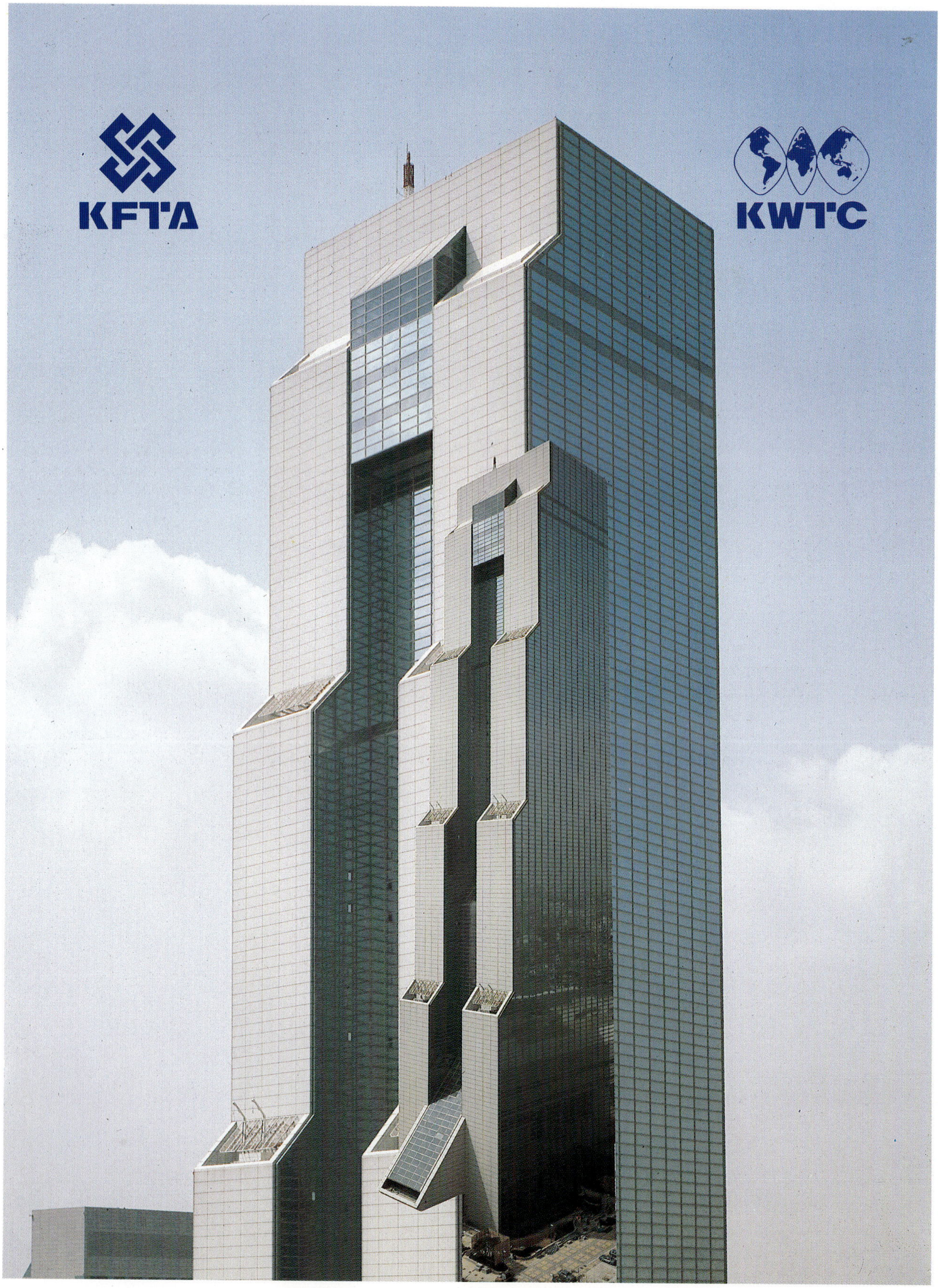
KFTA
KWTC

The Korea Foreign Trade Association

Korea World Trade Centre: a gateway to vast new trade possibilities

The Korea Foreign Trade Association (KFTA), founded in 1946 as a non-profit, private organization, is a full service trade organization of approximately 500 staff members with overseas branches in New York, Washington, Tokyo, Brussels, Dusseldorf, and Hong Kong.

The KFTA provides a wide range of business services involving the promotion of foreign trade. A substantial portion of the KFTA's services is devoted to trade information and international activities.

During the past decade, the KFTA has established a pattern of close cooperation and communication with domestic and foreign trade-related organizations for these purposes. The organization, which includes all importers and exporters in Korea, currently has a membership of over 14,000 companies. The KFTA has been a major force propelling Korea to the front-rank of the world's trading nations.

To facilitate international trade for domestic firms and assist foreign businessmen seeking business contacts with Korean partners, the KFTA established the Korea World Trade Centre (KWTC), which in close cooperation with world trade centres in other countries, provides visiting foreign businessmen with complete facilities for their business activities in Korea. These facilities include an exhibition building, trade information service, conference rooms and a hotel, as well as a shopping centre and city air terminal.

The world trade centre concept was conceived over 19 years ago in response to the rapidly growing volume of world trade. Since then many countries have established world trade centres on their own or with the assistance of the World Trade Centres Association (WCTA). Korea joined the WCTA in 1972 and as a member of the WCTA, maintains relations with over 150 organizations in more than 50 countries: from New York to Melbourne, Moscow to Rio de Janeiro, and London to Tokyo.

The KWTC: Korea's trade information centre

The KWTC, a three-year project started in March 1985 and completed in 1988, is Korea's first integrated trade complex and one of the largest trade centres in the world. It is located in Seoul's fast-growing Yongdong area, on a 200,000 square metre site that includes the most sophisticated exhibition facilities. The trade complex, which serves as a one-stop service centre for visiting foreign bussinessmen as well as Korean traders, consists of a 55-storey main office building, a four-storey exhibition building, 33-storey hotel, shopping centre, city air terminal, and convention centre. The US$580 million centre was inaugurated in August 1988, offering special support to small and medium businesses and providing a one-stop service that includes everything from trade administration procedures to trade enquiries as well as all other trade information services required by foreign and Korean businessmen. As an international information centre, and as a meeting place for

Duck-Woo Nam, Chairman of Korea Foreign Trade Association (KFTA).

businessmen, it plays an important role in the promotion of international trade.

The new centre is designed to serve several purposes: to meet the growing need for a centralized trade information facility; to offer visiting buyers every facility and convenience, including accomodation, recreation, leisure-sports and shopping, together with a one-stop service for business contracts and negotiations; to provide specialized support facilities for small businesses. The project will be a gateway to vast new trade possibilities. It is an important Korean achievement to be shared with the rest of the world.

The Trade Tower

For one-stop services, the Trade Tower accomodates all trade-related organizations, agencies and many individual firms, including banking, insurance and shipping companies. This building is equipped with telex, facsimile, electronic mailbox, and LAN (local area network) facilities for the efficient exchange of trade information. Centralizing these organizations and facilities makes this project even more cost-effective and efficient.

Exhibition building

Since foreign buyers are reluctant to make their purchases from catalogues and brochures, and prefer to see what they are buying, there is a growing need for exporting countries to provide a shop window for manufacturers. The new four-storey exhibition building and its three-storey annex, managed by the Korea Exhibition

Centre (KOEX) under the wing of the KFTA, are designed to meet this need in Korea.

The exhibition building, with a floor space of 200,000 square metres including the annex, is equipped with the most modern, functional facilities for the maximum convenience of exhibitors and visitors. All exhibitions staged here will be designed for selling rather than just advertising.

KFTA-Senior Management for International Affairs: From right: Vice-Chairman, Sun-Ki Lee; Managing Director, Nam-Hong Cho; Senior Managing Director, Chin-Shik Noh

As the core facility of the KWTC, the exhibition building is designed to accomodate both trade shows and permanent exhibitions in its three large exhibition and display halls and the three-storey annex.

The exhibition halls are furnished with the best and latest facilities and technologies and can host about 40 international exhibitions and fairs annually.

In the display halls, visiting traders will have the opportunity to see displays of Korean export and import products. The permanent exhibition booths in each display hall are staffed by company representatives to answer queries and provide information. Glass-enclosed sample showcases are provided to display major Korean export products.

At KOEX, a variety of business support services are offered to make business transactions simple, convenient and enjoyable. To assist business visitors, consultants and company representatives will be available to promptly answer enquiries and provide information. For the convenience of trade visitors, contracts can be arranged on the spot, so that they can have enough time to attend to other important buusiness or to enjoy seeing the sights of Seoul.

Exhibition halls

The Pacific, Atlantic and Continental halls in the exhibition building will be the main sites for international trade fairs and exhibitions. For maximum display flexibility and convenience, all three halls are equipped with sliding and folding walls.

The exhibition area will be designated a bonded area so that foreign visitors will be able to carry their exhibits into and out of the area without customs clearance difficulties. Other events such as conferences, conventions, seminars and theatrical performances will be held at the Olympia Hall on the third floor of the exhibition building. This hall is a spacious auditorium with a high-domed ceiling and is fully equipped with the most advanced audio and visual facilities. It can hold approximately 5,300 at one time.

Display halls

The display halls in the exhibition building are divided into two parts: the trade mart and the sample showcase.

The trade mart, composed of the export mart, import mart and the company showrooms, will display a wide range of Korean export and import products. The company showrooms feature information on major Korean companies and their products.

An estimated 800 permanent exhibition booths will be staffed by company representatives to assist visiting traders and provide detailed information on all products on display.

The sample showcase will show the quality of export products made by small and medium Korean enterprises. In all, there are approximately 700 sample showcases where consultants are readily available to answer queries or arrange transactions.

Business support services and facilities

Visiting traders will have immediate access to numerous business support services and facilities. Among the services to be provided will be telecommunications, simultaneous translation, audio and video services, on-line terminals for access to goods, exhibits and trade matters, and secretarial and interpretation services. Business support facilities will include banks, trade counselling offices, a customs office, post office, car rental offices, hotel reservation

*Trade Tower
*Hotel
*Shopping Center
*City Air Terminal
*Convention Center
*Exhibition Building

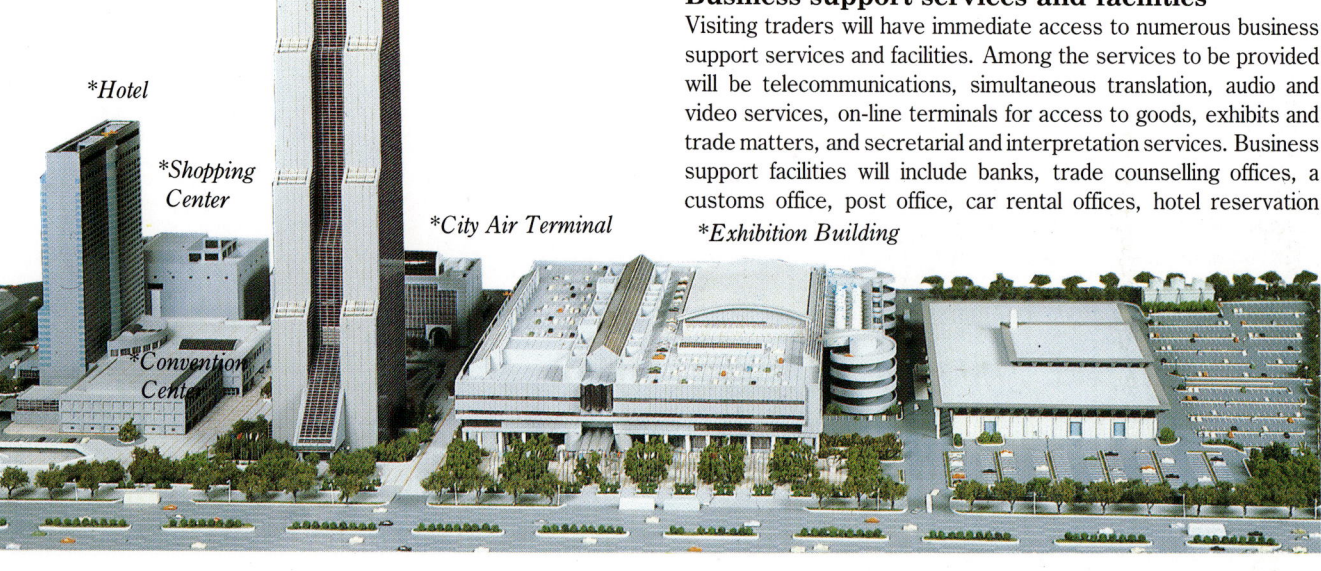

counters, tourist information counters, and catering and dining facilities.

Hotel

The new 33-storey hotel with 602 guest rooms and suites is conveniently located in the trade complex for foreign tourists as well as trade buyers visiting Seoul. Trade buyers in particular will find that the convenient location of the hotel will greatly enhance their access to Korean company representatives as well as various trade fairs and exhibitions held at KOEX. By staying at the Hotel Inter-Continental Seoul is an impressive new addition to Inter-Continental's worldwide chain of fine hotels. All of its guest rooms feature contemporary luxury complemented by traditional Korean art.

KFTA computer Room

Convention centre

The modern and fully equipped convention and conference facilities are a feature of the KWTC. From the vast Olympia Hall in the exhibition complex to the wide range of large, medium and small-sized rooms in the Trade Tower and hotel, the KWTC is destined to become the nation's premier meeting and convention site.

Shopping centre

The shopping centre is a part of Seoul's new decentralization plan. It enables visitors to the KWTC and participants in international conferences to easily combine their business and shopping activities.

City air terminal

The city air terminal includes an immigration office, quarantine station, customs house and travel agency. Transportation service to and from the airport is also provided, making it possible for trade visitors and tourists alike to minimize the time required for dealing with immigration procedures.

The KFTA's business services: national and international

The KFTA offers the businessman a wide variety of trade information and other services such as:

- Trade enquiry services for foreign businessmen interested in trading with Korea. The KFTA provides assistance to trade visitors in preparing their itinerary and identifying firms and individual businessmen to contact. If a personal visit is not possible, enquiries by telephone or letter will receive prompt attention. Questions concerning both export to and import from Korea are promptly answered by return post. Foreign businessmen are cordially invited to use the Trade Enquiry Office at the KFTA headquarters where qualified staff are always available.

- The EDPS-based data service, which compiles and analyzes various trade-related data and statistics for private businesses and government agencies. KFTA is the only private organization in Korea authorized to compile official trade statistics.

KFTA publications

- Publication of trade literature:
Korea Trading Post, a twice-monthly products newsletter, provides updates on new Korean products and developments in industry. This publication, which includes the listings of Korean suppliers of those products, is a good source of information for foreign buyers and is circulated to businessmen all over the world.
Korea Export, a catalogue of major Korean export products, published in colour twice a year for distribution to foreign buyers.
Korean Trade Directory, lists the names, mailing and cable addresses, and products of Korean manufacturers, exproters, importers and trade associations.

- International communication services, telex and electronic mailbox. The electronic mailbox service enables users to publicize their business activities through the WTC network, identify prospective trading partners, and communicate with each other through the network at attractive rates. The KFTA began to operate the electronic mail system in 1984 in conjunction with the WCTA and IP Sharp Associates.

- Individual consultancy, guidance and assistance for small and medium businesses.

- Language training programmes with an emphasis on business terminology. Individual and group courses are available in English and Japanese.

- Organization of seminars and study conferences.

- KWTC club and restaurant, which is not only a meeting place for the business community but also offers catering services for members' private functions.

The Lucky Securities Company Limited

The Lucky Securities Company Limited, established in 1969 as a member of the Lucky-Goldstar Group, is one of the most advanced securities firms in Korea with a paid-in capital of over US$130 million and about 1,200 highly motivated professionals. This expertise is put to work throughout their network of 28 fully computerized branch offices throughout Korea and the overseas offices in New York, London, and Tokyo.

Lucky Securities, a member of the Korea Stock Exchange and the Korean Securities Dealers Association, plays a vital role in the Korean capital market by providing its clients with a comprehensive range of financial services. From its main business which includes brokerage, dealing, managing and underwriting of public and corporate securities, to its handling of short-term financing instruments such as commercial paper, certificates of deposit, and beneficiary certificates, etc., Lucky Securities is at the cutting edge of the development of the Korean capital market.

The company's commitment to enhancing the scope and quality of its services through facility expansion and computerization, brought about its outstanding business performance for the fiscal year 1987. The company handled US$9.7 billion in securities transactions and recorded a net income of US$27.9 million, representing remarkable increases of 158% and 174% respectively, over the previous year. Lucky Securities established the Lucky-Goldstar Economic Research Institute in 1986 and founded the Lucky Investment Management Company Limited in April 1988. The company is now able to meet the needs of its clients around the world better than ever before with analytical and timely investment advisory services.

Financial highlights of Lucky Securities (US$ '000s)

Fiscal year	1984	1985	1986	1987
Total assets	308,574	448,935	792,997	2,047,999
Shareholders' equity	25,928	29,759	66,655	198,613
Total revenues	57,021	43,974	80,188	142,746
Net income	3,003	4,760	10,181	27,936

Note: Fiscal year ends in March

The night view of Lucky-Goldstar's Twin Towers

Kun-Joong Lee, President & Chief Executive Officer

Why such keen foreign interest in the Korean capital market?

During the 1980s, the Korean capital market grew rapidly in volume and quality. With more than 430 companies listed on the Korea Stock Exchange and a total market capitalization of over US$60 billion, the market is now the 16th largest in the world and, more importantly, one of the most promising. Reflecting the strongly bullish mood of the market, the Korea Composite Stock Price Index stood at 721.08 on July 30 1988, up 339% from 163.97 at the end of 1985.

This remarkable growth should be attributed not only to the robust economy but the expected full liberalization of the market. The beginning of the market's internationalization can be traced back to the early 1980s when the Korean government announced a series of steps to be taken to carry out its capital market liberalization programme. The cornerstone of this policy emphasized the inflow of foreign capital in the first stage and promoted the outflow of domestic capital in the final stage. Since then, the internationalization of the market has gradually progressed towards its long-range goal.

By the end of 1987 there were several investment vehicles that enabled foreigners to invest indirectly in Korean securities. These include five open-ended on-shore investment trusts: Korea International Trust, Korea Trust, Seoul International Trust, Seoul Trust and Korea Growth Trust. There are also two close-ended offshore funds: the Korea Fund and the Korea Europe Fund devoted exclusively to North American and European investors, respectively. In addition there are four convertible bonds that have been issued in the Euromarket by four of Korea's corporate leaders: Samsung Electronics, Yukong, Daewoo Heavy Industries and Goldstar.

Furthermore, as the Korean economy plays a larger role in the world, the government will intensify its efforts to internationalize the Korean capital market. This is particularly true with the advent of Korea's substantial current account surplus. The government has permitted the conversion of the three outstanding overseas CBs into equity stocks in addition to the issuance of more overseas securities by Korean companies. The second tranche of the Korea Europe Fund was also launched in 1988. It is particularly noteworthy that Korean securities firms have been authorized to open foreign exchange accounts permitting them to hold foreign assets worth up to US$30 million. All of these developments have occurred in line with the government's plan to enhance the international competitiveness of the domestic securities industry as well as to promote overseas investment.

A front runner in international corporate finance in Korea

In order to meet the challenges of the local market liberalization as well as the trend toward securitization worldwide, Lucky Securities has been emphasizing its international business. The company is well known for its outstanding corporate finance service capability through its participation in many international security syndications. Lucky Securities has played a major role whenever Korean equity-related issues have been issued in overseas markets.

Major participation in the international underwriting business

Issue	Date	Total amount	Role
Korea Growth Trust	85.5	US$30 mil.	Co-Lead Manager
Goldstar CB	87.8	US$30 mil.	"
Seoul International Trust	85.4	US$30 mil.	Co-Manager
Seoul Trust	85.4	US$30 mil.	"
Korea-Europe Fund	87.4	US$30 mil.	"
Tateho Chem. BW	87.5	US$50 mil.	"
KEPCO Yen Bond	87.6	Y7.5 bil.	"
Cosmo Securities BW	87.6	US$50 mil.	"
Toyo Wharf BW	88.2	US$50 mil.	"
Nikko Securities BW	88.6	£50 mil.	"
Korea-Europe Fund	88.7	US$30 mil.	"
Korea Fund	84.8	US$60 mil.	Underwriter
EXIM Bank DM Bond	85.6	DM100 mil.	"
City of Pusan Yen Bond	85.10	Y20 bil.	"
Samsung Electronics CB	85.12	US$20 mil.	"
KEB Yen Bond	86.1	Y20 bil.	"
EXIM Bank DM Bond	86.3	DM100 mil.	"
CPA Right Issue	86.4	HK$1.5 bil.	"
Daewoo Heavy Ind. CB	86.5	US$40 mil.	"
KEPCO Yen Bond	86.5	Y20 bil.	"
KDB DM Bond	86.6	DM100 mil.	"
Korea Fund	86.6	US$40 mil.	"
Yukong CB	86.7	US$20 mil.	"
UAL Right Issue	87.4	US$282 mil.	"
Sumitomo BW	87.5	US$600 mil.	"
Scudder New Asia Fund	87.6	US$60 mil.	"
Brazil Fund	88.4	US$150 mil.	"
Thai Investment Fund	88.4	US$31.2 mil.	"
Thai Euro Fund	88.5	US$75 mil.	"
Taiwan Fund	88.5	US$25 mil.	"
Spain Fund	88.6	US$60 mil.	"
C. Itoh BW	88.7	US$400 mil.	"

In 1987, Lucky Securities sucessfully co-lead managed Goldstar's US$30 million convertible bonds issued in the Euromarket. These were the only Korean overseas convertible bonds to be floated in 1987 and was selected as the best issue of that year by Euromoney. As a result of this success, the company has been able to increase its international prestige as well as to enhance its ability to issue overseas securities independently. Without even mentioning Korea-related securities, Lucky Securities has also been active in joining syndicates for overseas issues by major foreign companies. This is a clear demonstration of the company's rapidly growing capability in the international corporate finance business.

Since 1986, Lucky Securities has sponsored the annual European Capital Market Seminar, which is organized by the Korean Listed Companies Association. The main objective of the seminar is to assist Korean listed companies to diversify their money supply source so that they might find low-cost financing by introducing them to the European capital market. The second seminar of 1987, in which nine firms and four press agencies from Korea participated, was held in cooperation with world-famous financial institutions in major financial centres including London, Zurich, and Frankfurt. Satisfactory appraisal of the seminars came from the domestic firms and the local institutions involved. Both parties are anxiously looking forward to future European Capital Market Seminars.

Investment information and trading service

Lucky Securities has committed itself to supplying its clients with accurate and timely investment services. In order to carry out these objectives in a rapidly changing financial environment, the

The fully-computerized data bank system helps Lucky Securities supply its clients with timely investment information.

company has invested heavily in the latest technology to offer the highest quality data processing and delivery system.

The company has a comprehensive information network employing the largest data processing service company in the country, STM (a joint venture company with EDS of GM and the Lucky-Goldstar group).

Lucky Securities, equipped with the Reuter Monitor and teleservice from APDJ, inputs major economic and corporate news items affecting the Korean market as well as daily stock prices of the Korea Stock Exchange into the monitor, so that foreign investors around the world can better keep abreast of the Korean capital market. The company is also planning to integrate the terminals of overseas offices with the data bank of its main office in order to provide more data and information for its overseas clients.

With the exception of its participation in the Korea Fund and the Korea Europe Fund, international brokerage business by Korean Securities firms has been relatively inactive. However, this situation will change dramatically with international brokerage becoming the core of Lucky's Securities' activities once the Korean capital market is liberalized. Lucky Securities is positioning itself to become the most respected and dependable brokerage house in Korea by capitalizing on its professional expertise.

Towards the future

As the process of liberalization in the Korean capital market continues, Lucky Securities is ready to meet the challenges facing the securities industry and the new needs of its clients. To strengthen its ability to compete on a global scale, the company will continue to emphasize the expansion of its international presence.

Offices will be opened in financial centres such as Hong Kong, Zurich, and Frankfurt in addition to the existing offices in New York, London and Tokyo. These offices will be upgraded to branch offices or full subsidiaries. Lucky Securities is counting on this strategy of globalization to benefit its clients as well as to establish itself as a successful international force.

The roadshow of Goldstar CB Issuance.

Professional staff of Lucky Securities will meet the needs from its clients around the world.

Lucky Investment Management Company Limited

Lucky Investment Management Company Limited (LIMCO) was established on March 26 1988 as a wholly-owned subsidiary of The Lucky Securities Company Limited. LIMCO will strive to become Korea's most reliable adviser by capitalizing on its accumulated experience as an adviser to the Korea Europe Fund, the only close-ended Korean equity fund in Europe. LIMCO's primary activities include research into domestic and overseas economies, industries and companies in order to provide its clients with sound investment advice. Its strengths include a state-of-the-art information management system and an extensive database of listed Korean companies. Its staff is composed of top-class research personnel and fund managers all of whom have proven track records.

LIMCO's major activities in brief:

o Investment counselling and portfolio management
o Researching domestic and major overseas economies, industries and companies and providing its clients with the results
o Financial analysis of all listed Korean companies
o Fund management consultation in collaboration with The Lucky Securities Company Limited.

The Lucky Securities Company Limited

Lucky Investment Management Company Limited

Head office

34-6, Yoido-Dong, Youngdungpo-Ku, Seoul 150-010, Korea
Tel: (02) 784-7111, 7751, 7761
Tlx: LUSCO K 29771
Fax: (02) 784-8621
Reuter Code: LUCK-Y

London office

2nd Floor, 15 St Helen's Place, Bishopsgate,
London EC3A 6DE, United Kingdom
Tel: (01) 374-4812
Tlx: 918674 LUSCO LG
Fax: (01) 374-8350

New York office

One World Trade Center, Suite 1453, New York,
NY 10048, United States
Tel: (212) 432-7660, 7670
Tlx: 262220 LUSCO UR
Fax: (212) 432-7170

Tokyo office

4th Floor, Yusho Building, 2-7-2 Kayabacho Nihonbashi,
Chuo-ku, Tokyo, Japan
Tel: (03) 663-4281
Fax: (03) 663-4283

Merrill Lynch Capital Markets

Merrill Lynch established a representative office in Seoul in 1975 and we are currently among the most active of all investment banking firms in serving the corporate finance requirements of Korean banks and corporations worldwide.

Our leadership in Korea derives from several factors:
- We provide a wide range of investment banking service to our Korean clients,
- Our bankers in Seoul, Hong Kong, London and New York all have extensive experience in the Korean market. Most of our bankers covering Korea are fluent in Korean and have either lived or worked in Korea, and
- As a firm, we are familiar with the needs of our Korean clients and we have demonstrated our ability to successfully execute transactions on behalf of our Korean clients in all of the world's major capital markets.

Many of the transactions which we have arranged for Korean clients have involved innovative financing techniques or have been among the first transactions of their kind in Korea. For example, in 1976, Merrill Lynch lead managed a US dollar bond issue for Korea Development Bank. Three years later, we arranged a Kuwaiti Dinar denominated bond issue for the same issuer.

In 1984, Merrill Lynch arranged for the first Euro-commercial paper transaction by a Korean public sector issuer in the form of a US$75 million Revolving Underwriting Facility for Korea Exchange Bank. In 1986, we arranged and we presently serve as dealer for a US$150 million Euro-Commercial Paper Program also for Korea Exchange Bank. Our expertise in serving the short-term financing needs of our clients extends to the private sector as well. In 1985, we structured and arranged the first Euro-commercial paper program by a Korean corporate issuer in the form of a US$35 million Revolving Underwriting facility for Daewoo Corporation. These early transactions for Korean issuers in the Euro-commercial paper and Eurobond markets have helped pave the way for issues by other Korean private and public sector entities in these important securities markets. Likewise, Merrill Lynch's London-based traders are widely regarded as among the most knowledgeable and aggressive market makers in Korean short-term Euro-commercial paper today.

In the US commercial paper market, where Merrill Lynch is the industry leader, we have arranged and we currently serve as dealer on US commercial paper programs for virtually every major Korean private sector company. Transactions which we have structured and arranged include programs for Ssangyong Corp, Hyundai Auto Canada, Lucky-Goldstar International, Hyosung Corp. and Samsung Electronics America.

In addition to short-term financing transactions in the public markets, we have also arranged medium-term private placements for various Korean corporate entitites including Daelim Industries Ltd and Daewoo Corporation. Such transactions often incorporate interest rate or currency swaps from nearby transaction desks in Tokyo or Singapore.

In the area of equity and equity-linked financing, Merrill Lynch has played a pioneering role ever since the the Korean government announced initial guidelines in 1978 to allow for foreign investment in the Korean equity market. Merrill Lynch established one of the first vehicles for foreign portfolio investment in the Korean stock market — The Korea Trust — in 1981. Later, in 1984, we arranged and lead managed the second tranche of the Korea Trust. In 1985 we participated in the establishment of the Seoul Trust and in 1986, we acted as Co-lead Manager for the Korea Emerging Companies Trust which was one of the first Korean venture capital funds designed for placement with foreign investors.

In November 1985, the Korean government published new guidelines permitting Korean corporations to issue convertible bonds and similar equity and equity-linked securities abroad. Merrill Lynch has played an active role in virtually all such issues. We served as co-manager for the Samsung Electronics convertible bond issue in December 1985 and in the Daewoo Heavy Industry convertible bond issue in May 1986. In August 1987, Merrill Lynch lead managed a US$30 million offering of convertible bonds for the Goldstar Co. Ltd. The terms of this issue were among the most attractive ever obtained by a Korean corporate issuer.

As one of the largest securities firms in the world and a leading underwriter of equity offerings worldwide, Merill Lynch is well positioned to serve the needs of Korean companies in the international equity markets. Through our extensive network of sales offices and financial consultants, we are able to offer widespread distribution to both institutional and retail investors in all parts of the world. Widespread distribution, together with high quality country and company research and a strong commitment to secondary market trading have distinguished Merrill Lynch as a leader in international equity transactions for Korean corporations.

Most recently, as Korean firms have continued to expand into foreign markets, Merrill Lynch has emerged as a leader in the field of Merger and Acquisition advisory services. As more Korean companies seek to acquire strategic stakes in US or European firms whose businesses are complementary to their own, Merrill Lynch's skilled team of M&A professionals will be available to meet the specialized needs of Korean clients in this area.

Briefly, our principal investment banking services in Korean include:

- Merger and Acquisition Advisory Services
- Euro-Convertible Bond Issues
- US Commercial Paper Facilities
- Euro-Commercial Paper Facilities
- Private Placement Services
- Syndicated RUF and Loan Products
- Ratings Advisory Services
- Aircraft and Equipment Leasing
- Currency and Interest Rate Swaps
- Investment Management Services

Merrill Lynch stands ready to serve the changing needs of Korean public and private sector entities worldwide. Some of our most noteworthy transactions are highlighted in the following pages.

For additional information, contact:

Mr John Wisniewski
Chief Representative
Merill Lynch International Inc
Kyobo Building
Seoul, Korea

Tel: 82-2-735-7651

December 1976

THE KOREA DEVELOPMENT BANK

U.S. $25,000,000

9½% Bonds Due 1981

December 1981

The Korea Trust

U.S. $15,000,000

Issue of Units for Subscription

December 1982

Ssangyong (USA) Inc.

U.S. $40,000,000

Commercial Paper Program

September 1984

KOREA EXCHANGE BANK

U.S. $75,000,000

Revolving Underwriting Facility

July 1985

DAEWOO

Daewoo Corporation

U.S. $35,000,000

Revolving Underwriting Facility

June 1986

DAELIM

Daelim Industrial Co., Ltd.

Swiss Francs 17,820,000

Medium Term Private Placement

August 1986

The Export-Import Bank of Korea

U.S. $50,000,000

Interest Rate Swap Facility

August 1986

SAMSUNG ELECTRONICS AMERICA INC.

$50,000,000

Letter of Credit Facility

March 1987

HYUNDAI
AUTO CANADA INC

U.S. $50,000,000

Letter of Credit Facility

July 1987

GoldStar
GOLDSTAR CO., LTD.

U.S. $30,000,000

1¾% Convertible Bonds Due 2002

August 1987

Hyosung (America), Inc.

U.S. $40,000,000

Commercial Paper Program

April 1988

LUCKY·GOLDSTAR

Lucky-Goldstar
International (America), Inc.

$45,000,000

U.S. Commercial Paper Program

Acknowledgements

The photographs in the book were provided by: Colorific Photographic Library, Frank Spooner Pictures, The Hutchison Library, The Image Bank, Magnum Photos, The Robert Harding Picture Library and Tropix Photo Library.

The charts were drawn by Multiplex Techniques Ltd.

The author dedicates this book to his wife Nan. Without her support, editorial assistance and encouragement, this book would not have been written.